Lab Manual to Accompany

Soil Science & MANAGEMENT

6TH EDITION

Join us on the web at
www.cengage.com/community/agriculture

Lab Manual
to Accompany

Soil Science & Management

6TH EDITION

Edward J. Plaster

Developed by
Philip G. Gibson

DELMAR
CENGAGE Learning

Australia • Brazil • Japan • Korea • Mexico • Singapore • Spain • United Kingdom • United States

Soil Science and Management, 6th Ed.
Edward J. Plaster, M.Ed.
Philip G. Gibson

Vice President, Careers & Computing:
  Dave Garza

Senior Acquisitions Editor: Sherry Dickinson

Director, Development, Careers & Computing:
  Marah Bellegarde

Senior Product Manager: Laura J. Stewart

Editorial Assistant: Scott Royael

Vice President, Marketing: Jennifer Ann Baker

Director, Market Development: Deborah Yarnell

Senior Market Development Manager:
  Erin Brennan

Senior Brand Manager: Kristin McNary

Senior Production Director: Wendy A. Troeger

Production Manager: Mark Bernard

Senior Content Project Manager:
  Elizabeth C. Hough

Senior Art Director: David Arsenault

Media Editor: Debra Bordeaux

© 2014, 2009, 2003, 1997, 1991 Delmar, Cengage Learning

ALL RIGHTS RESERVED. No part of this work covered by the copyright herein may be reproduced, transmitted, stored, or used in any form or by any means graphic, electronic, or mechanical, including but not limited to photocopying, recording, scanning, digitizing, taping, Web distribution, information networks, or information storage and retrieval systems, except as permitted under Section 107 or 108 of the 1976 United States Copyright Act, without the prior written permission of the publisher.

> For product information and technology assistance, contact us at
> **Cengage Learning Customer & Sales Support, 1-800-354-9706**
> For permission to use material from this text or product,
> submit all requests online at **www.cengage.com/permissions**
> Further permissions questions can be emailed to
> **permissionrequest@cengage.com**

Library of Congress Control Number: 2012938489

ISBN-13: 978-0-8400-2434-3

ISBN-10: 0-8400-2434-7

Delmar
Executive Woods
5 Maxwell Drive
Clifton Park, NY 12065
USA

Cengage Learning is a leading provider of customized learning solutions with office locations around the globe, including Singapore, the United Kingdom, Australia, Mexico, Brazil, and Japan. Locate your local office at **www.cengage.com/global**

Cengage Learning products are represented in Canada by Nelson Education, Ltd.

To learn more about Delmar, visit **www.cengage.com/delmar**

Purchase any of our products at your local college store or at our preferred online store **www.cengagebrain.com**

Notice to the Reader
Publisher does not warrant or guarantee any of the products described herein or perform any independent analysis in connection with any of the product information contained herein. Publisher does not assume, and expressly disclaims, any obligation to obtain and include information other than that provided to it by the manufacturer. The reader is expressly warned to consider and adopt all safety precautions that might be indicated by the activities described herein and to avoid all potential hazards. By following the instructions contained herein, the reader willingly assumes all risks in connection with such instructions. The publisher makes no representations or warranties of any kind, including but not limited to, the warranties of fitness for particular purpose or merchantability, nor are any such representations implied with respect to the material set forth herein, and the publisher takes no responsibility with respect to such material. The publisher shall not be liable for any special, consequential, or exemplary damages resulting, in whole or part, from the readers' use of, or reliance upon, this material.

Printed in the United States of America
2 3 4 5 6   25 24 23 22 21

CONTENTS

LIST OF FIGURES	VI
PREFACE	VIII
INTRODUCTION	IX
ABOUT THE AUTHOR	X
Chapter 1 | The Importance of Soil	1
Chapter 2 | Soil Origin and Development	5
Chapter 3 | Soil Classification and Survey	9
Chapter 4 | Physical Properties of Soil	13
Chapter 5 | Life in the Soil	18
Chapter 6 | Organic Matter	21
Chapter 7 | Soil Water	26
Chapter 8 | Water Conservation	30
Chapter 9 | Drainage and Irrigation	34
Chapter 10 | Soil Fertility	39
Chapter 11 | Soil pH and Salinity	43
Chapter 12 | Plant Nutrition	47
Chapter 13 | Soil Sampling and Testing	52
Chapter 14 | Fertilizers	59
Chapter 15 | Organic Amendments	65
Chapter 16 | Tillage and Cropping Systems	69
Chapter 17 | Horticultural Uses of Soil	73
Chapter 18 | Soil Conservation	81
Chapter 19 | Urban Soil	85
Chapter 20 | Government Agencies and Programs	90

LIST OF FIGURES

Figure 1-1.	Visual inspection of soil properties.	2
Figure 1-2.	Soil solution and buffering capacity.	3
Figure 1-3.	Structural support.	3
Figure 2-1.	Flooding the ice tray.	6
Figure 2-2.	Physical weathering.	6
Figure 2-3.	Chemical weathering.	7
Figure 3-1.	Soil survey.	11
Figure 4-1.	Soil textural triangle.	15
Figure 4-2.	Soil particle shape.	16
Figure 4-3.	Soil porosity.	16
Figure 5-1.	Microorganism inspection.	19
Figure 5-2.	Large organism survey.	20
Figure 6-1.	Organic matter content.	22
Figure 6-2.	Pot drainage.	23
Figure 6-3.	Organic matter and microbial activity.	24
Figure 6-4.	Organic matter and soluble salts.	24
Figure 7-1.	Double-ring infiltrometer.	27
Figure 7-2.	Infiltration rate.	27
Figure 7-3.	Percolation rate.	28
Figure 7-4.	Drainable pore space and available water held.	28
Figure 8-1.	Soil water-holding capacity.	31
Figure 8-2.	ET_{rate} and watering plants.	32
Figure 9-1.	Infiltration rate.	36
Figure 9-2.	Soil percolation columns.	36
Figure 9-3.	Percolation rate.	37
Figure 10-1.	Soil sample lab report.	40
Figure 10-2.	Recommendation for soil fertility management.	41
Figure 10-3.	Fertility program.	41
Figure 10-4.	Fertilization record.	42
Figure 11-1.	Soil pH and soluble salts.	44
Figure 11-2.	Marigold plant development utilizing varying watering solutions.	45
Figure 12-1.	Soil sample report.	48
Figure 12-2.	Plant nutrition strategy.	49
Figure 12-3.	Phosphorus availability.	49

Figure 12-4.	Corn crop nitrogen recommendation.	50
Figure 12-5.	Nitrogen application schedule for corn.	50
Figure 12-6.	Nutrient solution culture.	51
Figure 13-1.	Site planning information.	53
Figure 13-2.	Soil-sampling and testing plan.	54
Figure 13-3.	Residential site plan.	55
Figure 13-4.	Soil-sampling and testing plan.	56
Figure 13-5.	Soil-sampling and testing plan for soybean crop.	57
Figure 14-1.	Soil report.	61
Figure 14-2.	Nutrient source and blending recommendation.	62
Figure 14-3.	Fertilizer label.	62
Figure 14-4.	Fertilizer calculations.	63
Figure 14-5.	Granular fertilizer application pattern.	63
Figure 14-6.	Calibration graph.	64
Figure 15-1.	Chemical properties of soil, amendment, and amended soil.	66
Figure 15-2.	Physical properties of natural soil, soil–compost mixture, and soil–pine bark mixture.	67
Figure 16-1.	Rototilling effects on bulk density and porosity.	71
Figure 16-2.	Growth rate.	71
Figure 16-3.	Mulch effects.	72
Figure 17-1.	Recommendations for peach production in the southeastern United States.	75
Figure 17-2.	Soil management and modification program for peach production in the southeastern United States.	76
Figure 17-3.	Tomato production soil requirements and recommendations.	77
Figure 17-4.	Removal of one-half of the root ball with a spade.	78
Figure 17-5.	Observations of root growth and development after transplanting from the original root ball.	79
Figure 18-1.	Water erosion.	82
Figure 18-2.	Soil erosion prevention.	83
Figure 18-3.	Soil erosion prediction.	84
Figure 19-1.	Urban soil survey.	86
Figure 19-2.	Urban soil management plan.	87
Figure 19-3.	Soil comparisons.	88

PREFACE

The Lab Manual to accompany *Soil Science & Management,* 6th Edition, by Edward J. Plaster is a collection of activities designed to augment the principles taught in the text. The manual follows the text closely. There are 20 chapters that align with the text. Each chapter contains an introduction, equipment and materials list, and activities relating to the text material. Students can complete all or selected portions of each chapter's activities. In addition, many of the activities can be modified to allow students to complete them off-campus or at home. In fact, the lab activities can be easily adapted to an Internet-based soils course.

Soil science covers a wide range of applications, from soil engineering to golf course management. This Lab Manual provides a suitable introduction to all the various aspects of soil science. Although the primary focus is soil management relative to plant culture systems, soil engineering applications are also incorporated into several of the lab activities. An instructor can place variable emphasis on each chapter to facilitate appropriate learning based on the students' needs.

The act of discovery is the most effective learning tool in all disciplines. The use of this Lab Manual accentuates these "discovery" events for soil science students. Students should be encouraged to complete the activities with minimal direction from the instructor. The development of a clear understanding of soil science comes more from figuring things out than from following a recipe. In the lab, we are not baking bread; instead, we are learning how to discover.

NEW TO SIXTH EDITION OF THE LAB MANUAL

In general, content has been updated to correlate with the updates made to the core textbook. In particular, Chapter 1 incorporates updates to two activities. The third activity now relates to determining pH, and the last activity provides an updated activity related to the structural support of soil. Chapter 5 now includes a new Internet-based activity and corresponding Challenge Question related to conducting a survey of soil organism genetics. Chapters 7 and 19 contain new information related to water movement and the Soil Erosion and Sediment Control Act and LEED certification, respectively.

INTRODUCTION

Soil science is both an ancient and a modern topic. Soil management is a well-documented cultural practice in many civilizations dating as early as agriculture itself. Today, there is a resurgence of interest in soil science and management. Organic growing techniques capitalize on the effectiveness of careful soil management, and traditional growing practices are under increasing demands for conservation and productivity. Soil science students have a unique opportunity to learn and apply the soil management principles covered by this Lab Manual and its accompanying text.

This Lab Manual is intended to enable the soil science student to encounter soil closely. It begins with introductory exercises designed to familiarize students with the material we call *soil*. Next, the activities gradually increase in complexity and align with the principles covered by the accompanying text. Students are encouraged to choose the activities that most interest them. Modifications to the activities are also encouraged. Most important, students are encouraged to discover the principles underlying the activities. How do the results of each activity resolve the basic principles of soil science? Why would each of these resolutions be important in an effort to manage soil? These core questions should be the overriding purpose of each activity in this Lab Manual.

Soil science has a broad range of applications. Everything from structural support to cotton production relies heavily on a clear understanding of the soil. In turn, these applications rely exclusively on well-trained soil scientists and engineers who can make competent recommendations for soil management practices. This introductory Lab Manual provides the opportunity for students to establish a firm basis in soil science. A student can easily build on these initial discoveries to become an expert in any field relating to soil.

Students should plan on discovery as being the primary objective of the activities included in this manual. Much like learning to swim, where the basic principles of buoyancy become immediately apparent on diving into the water, the activities in this Lab Manual bring students into direct contact with the soil. If they can perceive the activities as an opportunity rather than a chore, the lesson is already half-learned.

ABOUT THE AUTHOR

Dr. Philip G. Gibson is the Bioscience Program Director at Gwinnett Technical College in Atlanta, Georgia. He holds a Ph.D. in Plant Physiology from Clemson University, an M.S. in Soil Physics from the University of Georgia, and a B.S. in Nursery Management from Oklahoma State University. Dr. Gibson has 24 years' experience teaching and 10 years' experience working in the horticulture industry.

Dr. Gibson and his wife have three wonderful children, who enjoy many outdoor activities together. An environmentalist at heart, the unquenchable desire to understand the natural world drives Dr. Gibson to teach, and learn from students in an educational forum.

CHAPTER 1

The Importance of Soil

PURPOSE

This chapter is designed to impress upon you the importance and properties of soil. Soil is a critical component of ecology, climate, architecture, engineering, resources, and life.

OVERVIEW

The soil is a complex organization teeming with ecology all its own. An entire lifetime could be spent identifying the unknown living organisms in the soil surrounding a single oak tree. Soil scientists dedicate their careers to understanding perhaps the most complex ecological system on the planet. One teaspoon of soil contains 12 million living microbes, in addition to a healthy population of nematodes, fungi, and plant roots. The mineral matter found in soils is composed of weathered rocks broken into tiny fragments. Some of these fragments are microscopic and together with organic matter serve as a reservoir of elemental nutrition for plants. This collection of soil particles that supports plant life can be characterized both physically and chemically. It is important to realize that these characterizations are separated only by analysis. The soil system includes both chemical and physical aspects that cannot be separated in a growing environment.

A preliminary examination is a crucial first step in understanding the various processes and properties of soil. Much can be learned by simply looking closely at what comprises a soil collected from a site of interest. Additionally, the amazing ability of soil to maintain a thriving ecosystem designed to provide for the needs of plants can be easily recognized by close observation. Soil protects this ecosystem by resisting change or *buffering* the soil environment against agents of change. These change agents can be chemical or physical, as is illustrated by the activities in this unit. Soil scientists and engineers use their knowledge of soil properties to capitalize on a soil's wide range of abilities, from an ideal environment for plant growth to basic structural support for architectural endeavors.

ACTIVITIES

Students are invited to encounter the components of natural soils and explore their various uses. Begin by collecting four soil samples from different locations. Also, try to find soil samples that appear and feel different. Collect approximately 1 cup (~350 g) of each sample and place them in separate small paper bags. Number the samples and record pertinent information about each sample (e.g., location collected, soil usage, soil history, etc.).

Equipment and Materials

soil collection paper bags
graduated cylinder (100 mL)
scale
10× field lens
distilled water
150-mL beaker or jar
plastic spoon

glass microscope slides
30× or higher microscope
0.5-mL eyedropper
vinegar
litmus paper

Activity

1. *Visual inspection.* Examine each of the four soil samples carefully by spreading out 3 or 4 spoonfuls on a clean paper towel. Feel the soil with your fingers. Observe the soil components using your naked eye and a 10× field lens. Make note of the color and texture of the soil. Is it red or gray, gritty or smooth? Does the soil contain living organisms, dead organisms? Are there plant roots? Is the soil moist or dry? Are the particles the same size or varied? Record your observations in Figure 1-1.

Sample Number	Naked Eye Inspection	10× Magnified Inspection
1		
2		
3		
4		

Figure 1-1
Visual inspection of soil properties.

2. *Soil solution inspection.* Mix 2 tablespoons of one of the soil samples with 2 tablespoons of water in a 150-mL beaker or jar. Stir thoroughly. Let stand for five minutes. Collect a small amount of the liquid soil solution with an eyedropper. Place a drop of the solution onto a glass microscope slide and observe the solution through a 30× or higher microscope. Do you see any living organisms? If so, how can you tell they are alive? What color is the soil solution? Repeat this procedure for the other three soil samples. Record your observations in Figure 1-2.

Sample Number	Soil Solution	Soil Solution pH with Vinegar
1		
2		
3		
4		

Figure 1-2
Soil solution and buffering capacity.

3. *Buffering capacity.* Determine the pH of vinegar with litmus paper. Mix 2 tablespoons of one of the soil samples with 2 tablespoons of vinegar in a 150-mL beaker or jar. Stir thoroughly. Let stand for five minutes. Collect a small amount of the liquid soil solution with an eyedropper. Test the solution pH with litmus paper. Repeat this procedure for the other three soil samples. Record your observations in Figure 1-2.

4. *Structural support.* Attempt to shape each of the four soil samples into a ball. Make note of the soil's tendency to retain its shape or its lack of ability to be molded. Wet each of the four soil samples with a small amount of water. Make note of the soil's texture and/or grittiness. Record your observations in Figure 1-3.

Sample Number	Ability to Be Molded	Grittiness
1		
2		
3		
4		

Figure 1-3
Structural support.

CHALLENGE QUESTIONS

1. What properties of soil can be easily seen?

2. What properties of soil cannot be easily seen?

3. Which properties of the soils you collected are similar?

4. Which properties of the soils you collected are different?

CHAPTER 2

Soil Origin and Development

PURPOSE

This chapter is designed to familiarize you with the language used by soil scientists and the processes by which soils are formed.

OVERVIEW

The term *soil* has a very broad application. Soil can refer to that layer of the earth that separates the atmosphere from the continental crust. Alternatively, soil can refer to that which makes up the ground under our feet. Jokingly, soil scientists separate the term *soil* from the term *dirt* by describing "dirt" as what you find on the bottom of your shoes or under your fingernails and "soil" as where plants grow. Many of the terms used by soil scientists serve to clarify exactly what is being meant by the term *soil*. Terms like *pedon, solum,* and *topsoil* are used to convey more specificity. Descriptive terms like *eolian, alluvial,* and *colluvial* provide further information about a soil or its properties. It is important for the student to learn these terms in order to discuss and understand various aspects of soil science and management.

Soil is made up of degraded rocks, formerly living material, and living organisms. Additionally, the spaces between these soil components play a critical role in the properties of a particular soil. The formation of soil is a rather lengthy process, but the factors contributing to soil formation can be easily mimicked in the laboratory. Physical and chemical weathering of parent material that occurs over long periods of time during soil formation determines the mineral characteristics of a soil. Organic materials, both living and dead, occurring in the soil are influenced by environmental conditions, as well as the mineral components and porosity of the soil. The activities in this chapter illustrate some of these processes in isolation. It is important to recognize that all these processes work in concert in nature, yielding the soils we find today.

ACTIVITIES

Soil formation results from a combination of factors, some of which are illustrated by these activities. Obtain samples of sand, silt, clay, and potting soil from the instructor. Conduct the following activities using these materials.

Equipment and Materials

sand (4 cups)
silt (1 cup)
clay (1 cup)
potting soil (3 cups)
ice cube tray

hydrochloric acid (10% to 20%; ~3 to 6 M)
running water
freezer
250-mL beaker or Mason jar
small paper cups (to hold one ice cube)

Activity

1. *Physical weathering.* Mix 1 cup of sand, 1/2 cup clay, and 1 cup of potting soil together. Pour the soil mixture into one-half of the ice cube tray. Mix equal volumes of sand, silt, and clay. Pour this soil mixture into the remaining half of the ice cube tray. Slowly run water onto the top edge of the ice tray as shown in Figure 2-1 until all of the slots of the tray are filled with water. Place the tray in the freezer overnight or until frozen solid. Remove the frozen soil samples from the freezer and eject them from the tray. Observe the profile of each cube. Allow the ice cubes to melt in separate cups and examine the resulting soil mixture. Record your observations in Figure 2-2.

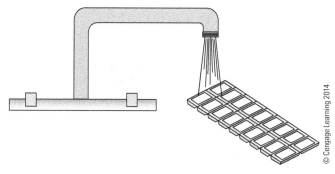

Figure 2-1
Flooding the ice tray.

FROZEN SOIL PROFILES	SOIL TEXTURE AND CONSISTENCY

Figure 2-2
Physical weathering.

2. *Chemical weathering.* Mix 2 cups of sand with 2 cups of potting soil. Fill the beaker or Mason jar partially full with half of the soil mixture. Retain the rest for comparison later. Have the instructor add hydrochloric acid in sufficient volume to slightly moisten all of the soil mixture. Use extreme caution since the acid is quite corrosive. Let stand for 72 hours or until dry. Add water in sufficient volume to thoroughly soak all of the soil mixture and let stand for 30 minutes. Drain the water from the soil mixture, retaining the soil with a paper towel or fingers. Allow the soil to air-dry or oven-dry the mixture. Compare the acid-treated soil mixture with the remaining portion of the original mixture. Record your observations in Figure 2-3.

ORIGINAL SOIL MIXTURE	ACID-TREATED SOIL

Figure 2-3
Chemical weathering.

CHALLENGE QUESTIONS

1. How does freezing and thawing work to form soils?

2. How would soils be exposed to acid in nature?

3. How do environmental conditions affect organic matter development?

4. What causes the formation of layers in the alluvium?

CHAPTER 3

Soil Classification and Survey

PURPOSE

This chapter is designed to illustrate the process of classifying soils and using such classifications to develop useful soil surveys.

OVERVIEW

Soil classification systems have been in use since the ancient Egyptians developed strategies for maximizing crop yields over 2,000 years ago. We still use similar types of strategies today for some of the same reasons. The United States Department of Agriculture (USDA) or the Natural Resources Conservation Service (NRCS) provide excellent resource information for growers or civil engineers needing soil survey information to develop soil utilization strategies. Soil classification information is valuable for many reasons, but often land capability classes or even more focused soil-use maps can be of greatest practical application. Bridging the gap between broad-based soil information and site-specific soil-use recommendations can be a challenge for the soil scientist. It is important to learn to apply broader concepts to specific situations.

Gathering soil information on a particular site is useful to planners and developers as they create land-use plans. This information can also be very useful to growers as they develop strategies for plant growth and development. Building locations, drainage patterns, planting plans, and various other land uses rely on accurate soil maps. A soil map can be developed for any sized plot of land. The soil sampling locations should be plotted on a scaled drawing. The degree of specificity of a soil map is determined by the intensity of land usage and the types of soil information required. Information that might be collected includes slope proportions, soil texture, soil structure, pH, nutrient level, profile, and organic matter content at several different locations. A soil scientist might make specific recommendations for soil modification or changes of grade on the site. As land use and the management of plants intensify, the development of soil maps will become increasingly desirable in the future.

ACTIVITIES

Students will develop a soil map for an area found on the campus or near the school. Students may choose a site, or the instructor can assign a site for students to utilize, to develop a soil map. Select a location that is large enough to have variable conditions but small enough to measure (e.g., 1/4 to 1/2 acre).

Equipment and Materials

18 in. × 24 in. vellum

pencils and erasers

straightedge

engineer's or architect's scale

soil-sampling probe

soil collection bags

tape measure

measuring wheel

compass

Internet access

Activity

1. *Collecting soil information.* Visit the Web site http://soils.usda.gov to collect information regarding the soils in your area. Visit the local NRCS office and the county extension agent's office to gather additional information about the soils in your area.

2. *Scaled drawing.* Prepare a scaled drawing or base map of the area to be mapped. *Scale* is the term used to describe the specific translation of the distances on a site into a smaller unit of measure to facilitate accurate drawings that represent the site. Use a scale appropriate for the area of the site (e.g., 1/8 in. = 1 ft, or 1 in. = 10 ft). Locate all the existing structures, drainage areas, vegetation, and other pertinent information on the drawing. Indicate the direction north on the drawing. Label the site location by title, street address, and date along the edge of the drawing.

3. *Land capability classes.* Determine the land capability class or classes associated with your site. Refer to your *Soil Science & Management* text or the NRCS Web site for more information on land capability classes.

4. *Soil sampling.* Collect at least 10 samples from different locations on the site. Number each sample (e.g., 1, 2, etc.). Indicate the sample collection location on your scaled drawing by labeling the sample number on the base map at the location it was collected. Describe the soils from each sampling location in your own words (e.g., light, dark, coarse, dry, etc.) and speculate on its potential use (e.g., lawn, roadway, trees, etc.) in Figure 3-1.

Sample Number	Soil Description	Potential Use
1		
2		
3		
4		
5		
6		
7		
8		
9		
10		

Figure 3-1
Soil survey.

CHALLENGE QUESTIONS

1. What types of soil are common to your area?

2. What are the more common parent materials for soils found in your area?

3. How could your soil map be used by a developer?

4. How variable is the soil on your soil map?

CHAPTER 4

Physical Properties of Soil

PURPOSE

The purpose of this chapter is to determine several physical properties of soil.

OVERVIEW

The most important of the soil characteristics are the physical properties. These properties have a greater influence on plant health and other soil uses than any other item. Physical properties of the soil include soil texture, soil structure, and organic matter content. The levels of moisture, oxygen, and mineral nutrients are determined by the physical properties of a soil. Fine-textured soils, those with higher proportions of small-sized particles, hold more moisture and nutrients than do coarse-textured soils. However, oxygen availability in fine-textured soils can limit plant growth and development. Coarse-textured soils have higher oxygen content, but may not hold sufficient amounts of moisture and nutrients. Soil structure describes the shape of soil particles and their arrangement. The structural characteristics of soil also affect cultural practices. Common problems that arise from soil structure are compaction, low infiltration or percolation rates, and perched water tables. Often, tilling is recommended to resolve structural problems. Porosity of soil is particularly influential in the growth and development of plants. The space between soil particles contains the water and oxygen critical to root growth. Additionally, the exchange of nutritional elements occurs between soil particles and water held in the soil pores. The volume and tortuosity of these soil pores determines the ease with which soil water and air can be refreshed. Plant culture programs should be sensitive to the soil's physical characteristics. For example, the frequency of fertilization and irrigation is dictated by soil physical properties. Therefore, the determination of a soil's physical properties should be incorporated into soil or plant management programs.

ACTIVITIES

Soil texture is best determined by a shaker test using sieves to separate the various particle sizes. However, soil texture can be estimated with several techniques, including the sedimentation test described here. Soil structure describes two different aspects of soil: the arrangement of soil particles and the shape of the individual particles. Both "soil structures" are important for soil management and plant culture. This activity is limited to the shape of individual particles. (The arrangement of soil particles is explored later.) Finally, the porosity of soil is relatively easy to determine but is absolutely critical to soil management programs.

Equipment and Materials

two large Mason jars

Borax

permanent marker

plastic spoon

50×–100× microscope

glass microscope slides

oven

paper towel

scale

3/4-in. pipe or soil-sampling probe

Activity

1. *Determine soil texture.* Collect a representative soil sample from a plant-growing area of your choice. What you are trying to accomplish is to acquire a sample that represents the root-growing environment of a group of plants that you intend to cultivate with the same program. This is different from a "random" sampling process. For example, if you are going to fertilize the shrubs in the front of your house all the same, you need a sample that represents an average of the soil in front of your house. Obviously, each of the plants in front of your house has a different nutritional need and is growing in a different soil condition, but it is not practical to develop a unique program for each plant. Similarly, it is not practical to develop a nutritional program for each child in an elementary school although, clearly, each child has differing nutritional requirements. To be practical and economical, the school system develops a lunch program that meets the nutritional needs of the average child (of course, if this does not meet the needs of your child, you have the option of preparing lunches yourself).

    Collect a sample from the root zone (0 to 6 in. below the surface) in at least 10 different locations around the plants. Make sure that you accumulate at least 2 cups of soil. Remove any undecomposed organic matter from your sample (e.g., roots, sticks, leaves, etc.). Place the sample in a Mason jar. Mix 1/2 teaspoon of Borax with 2 cups of water in a separate Mason jar. Stir until the Borax is completely dissolved. Slowly pour water into the jar containing the soil sample until there is 2/3 water and 1/3 soil. Wait 15 minutes. Then use a permanent marker to mark the level of soil in the jar. Measure the distance from the bottom of the jar to the mark (top of the soil) with a ruler and record the depth. Next, cover and shake the jar for 2 minutes (this is a long time, and your arm will be sore). Set the jar on a flat surface and measure the depth of soil that settles out in 30 seconds. This is sand. Next, measure the total depth of soil that settles out in 40 minutes. This is sand and silt combined. Use these measurements and the following information to determine the proportion of sand, silt, and clay in your sample:

    - % sand = (depth settling out after 30 seconds/total depth of soil at the start) × 100
    - % silt = [(depth settling out after 40 minutes − sand depth)/total depth of soil at the start] × 100
    - % clay = 100% − (% silt + % sand)

    Determine the soil texture by entering these proportions into an online soil calculator. Alternatively, use the textural triangle found in Figure 4-1 to determine your soil's texture. The soil textural classification is the textural "name" describing your soil (e.g., clay loam, loamy sand, silty clay loam, etc.). It can be determined by reviewing where your soil's sand, silt, and clay proportions fall on the textural triangle. Find where your sand and clay proportions intersect on the

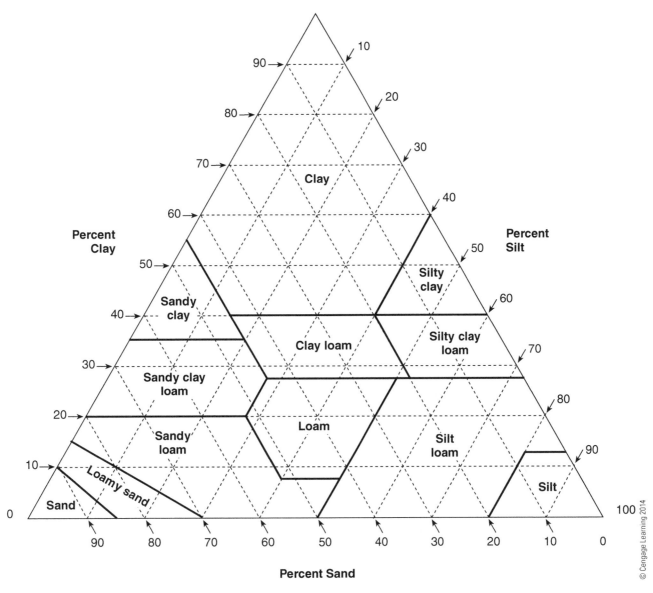

Figure 4-1
Soil textural triangle.

graph. Follow the graph key (L = loam, S = sand, etc.) to determine the textural classification of your soil.

2. *Determine soil particle shape.* Pour the excess water off the sample. Using a spoon, carefully extract some of the soil particles from each of the sedimentation layers. Place the soil particle extractions from each layer onto a separate glass microscope slide. Examine the particles under 50×–100× magnification with the microscope. Draw pictures of the particle shapes in Figure 4-2.

Sedimentation Layer	Particle Shape
Sand	
Silt	
Clay	

Figure 4-2
Soil particle shape.

3. *Determine soil porosity.* Collect soil cores using a hollow 3/4-in. pipe or a volumetric soil-sampling probe and the sampling procedure from activity 1. Label and place the sample on a paper towel in an oven. Make sure to include the volume, location, purpose, date, and texture of the soil on the label. Leave the sample at 105°C (221°F) in the oven for 72 hours. Remove the sample from the oven and weigh it. This can be accomplished by placing the paper towel with the soil core on the scale. Record the weight, remove the soil core, and record the weight of the paper towel. Deduct the weight of the paper towel from the total weight to determine the weight of your soil core. Finally, determine your soil's bulk density (BD) by dividing the weight (g) by the volume (cm^3).

Determine the soil porosity by utilizing your soil's BD and assuming the particle density (PD) to be 2.65 grams per cubic centimeter. This technique mathematically separates the volume occupied by the solids from the volume occupied by air (pore space) in your oven-dried soil sample. Comparing the volume of pore space to the total bulk volume of your soil describes its porosity. Complete these calculations for your soil sample in Figure 4-3.

$$\text{Porosity} = 100\% - (BD/PD \times 100)$$

$$\text{Porosity} = 100\% - (\underline{} / \underline{} \times 100)$$

$$\text{Porosity} = \underline{} \%$$

Figure 4-3
Soil porosity.

CHALLENGE QUESTIONS

1. Why isn't the addition of sand a practical way to change a fine soil to a coarser soil?

2. As the bulk density of a soil increases, what happens to the porosity?

3. What shape of soil particle would compact the easiest?

4. Which soil texture would have the greatest proportion of small pores or micropores? Which texture would have the lowest oxygen content?

CHAPTER 5

Life in the Soil

PURPOSE

This chapter is designed to explore the influence of living organisms in the soil and the fragility of soil ecology.

OVERVIEW

The ecology of soil has become a much more contemporary topic. Conservation movements, organic farming, and the Green Revolution have all brought the ecology of soil a newfound emphasis. A keen appreciation for the complexity of the rhizosphere has emerged as we learn more about the interaction between the environment, plants, and the soil. The array of microorganisms that coexist with larger organisms in the soil environment creates a system of checks and balances similar to what we normally associate with large-scale ecosystems like rainforests. An understanding of soil ecology enables soil scientists to more effectively manage and utilize the soil.

ACTIVITIES

Students are invited to extract and manipulate microorganisms living in natural and artificial soil environments.

Equipment and Materials

two small glass jars

distilled water

paper towels

glass microscope slides

50×, 100×, or 1,000× microscope

eyedropper

Activity

1. *Extraction of microorganisms.* Many soil microorganisms are so small that they are difficult to filter out of the soil. Collect 1 cup of a natural soil sample from a location of your choosing. Mix 1/2 cup of the soil sample with 3/4 cup of distilled water in a small jar. Stir the mixture until thoroughly mixed. Filter the soil: water mixture through a paper towel and collect the filtrate in a glass jar.

2. *Microorganism inspection.* Place a drop of the soil solution extract on a glass microscope slide, cover with a cover slip, and observe the microorganisms under 50×, 100×, and 1,000× magnifications. Record your observations in Figure 5-1.

Magnification	Soil Solution Extract
50×	
100×	
1,000×	

Figure 5-1
Microorganism inspection.

3. *Larger organism survey.* Dig around in several areas of soil under turfgrass or woodland. Capture and identify as many larger organisms (e.g., grubs, millipedes, sow bugs, worms, beetles) as possible. Characterize the organisms as beneficial or detrimental to plants. Enter your observations in Figure 5-2.

4. Conduct a survey of soil organism genetics by accessing GenBank at http://www.ncbi.nlm.nih.gov/genbank/ and entering the search term "Actinomyces" into the search window across the top. Review the results and select a few of the references to enter as Google searches.

Type of Organism	Detrimental	Beneficial
1		
2		
3		
4		
5		

Figure 5-2
Large organism survey.

CHALLENGE QUESTIONS

1. What types of microorganisms were present in your soil solution extract?

2. Were most of the larger organisms beneficial or detrimental? How could you manipulate this to your advantage?

3. What was found with the search of GenBank for soil-inhabiting organisms?

CHAPTER 6

Organic Matter

PURPOSE

The purposes of this chapter are to measure the amount of organic matter in a soil sample and to study the role of organic soil constituents.

OVERVIEW

The organic matter content of soil describes the amount of living and formerly living matter that exists in various stages of decomposition within the upper layers of soil. These particles shrink and swell with moisture content. This will help form large pores in the soil to improve aeration in fine-textured soils. Excessive amounts of organic matter can be problematic, as with histosols. Additionally, organic matter sticks small soil particles together to form aggregates, which further improves soil structure. It is important to recognize that the composition of the organic matter in a particular soil comes from a diverse group of living materials. In natural settings, organic matter composition in the soil is ideally suited to the ecosystem being supported by the soil. This interaction between life and death is exceedingly complex, as in allelopathic interactions. Much like natural ecosystems, a soil manager must determine the appropriate amount and kind of organic matter content that should be maintained in a soil. Such determination depends on what the specific goals are relative to a particular soil's usage. Strategies for soil organic matter management should be sensitive to the existing soil conditions, long-term environmental conditions, and the crop requirements. Green manure crops are sometimes recommended for soil-rebuilding activities directed toward plant growth and development over a one- to five-year period. Although time-consuming, green manure crops like sudan add significant quantities of rapidly decomposed organic matter to the soil. In other instances, bark products are recommended because of their resistance to decay. Bark products are more readily available and can be easier to apply than green manure crops.

Measuring the amount and type of organic matter found in a soil sample can require extensive lab procedures utilizing sensitive equipment. However, some simple procedures can be employed to get a general idea of organic matter content in a particular soil sample. These separation techniques utilize differences in degradation resistance between the organic matter and the mineral matter in a soil sample. Because organic matter is much less resistant to destruction by heat or chemical digestion, it can easily be removed from a soil sample. Once removed, weight or volume loss can be used to estimate the proportion of organic matter in a given soil sample. The proportion of organic matter usually ranges from 0.5 percent to 5.0 percent by volume. The overall proportion is useful data, but this provides little information about the kind of organic matter present or its contribution to the overall soil properties.

ACTIVITIES

Soil organic matter is removed from soil samples using various techniques. Upon removal, changes in weight or volume can be used to estimate the overall organic matter proportion.

Equipment and Materials

oven-dried soil samples
ceramic crucibles
crucible tongs
conductivity meter
kiln
scale
150-mL beaker
funnel
filter paper
glass stirring rod
distilled water
5.0-*M* sodium hydroxide

3.0-*M* hydrochloric acid
250-mL Erlenmeyer flask
six 4-in. plastic pots
peat
sand
three 20-mL collection vials
three glass microscope slides
1,000× microscope
bean seeds
three 5,000-mL beakers
pH meter

Activity

1. Determine organic matter content by ignition. Place 10 g of oven-dried soil into a crucible. Weigh the crucible and soil together. Place the crucible with soil sample into a kiln for two hours at 500°C. Remove the crucible with tongs and allow it to cool. Weigh the sample and crucible, subtract the weight of the crucible, and record your results in Figure 6-1.

Initial Sample Weight	Proportion of Organic Matter Determined by Ignition	Proportion of Organic Matter Determined by Chemical Digestion
(25 g)	100% − (10 g − _____ g)/10 g	100% − (25 g − _____ g)/25 g

Figure 6-1
Organic matter content.

2. Determine organic matter content by chemical digestion. Place 25 g of oven-dried soil into a 150-mL beaker. Carefully add 20 mL of 5.0-*M* sodium hydroxide (pH 13.0). Stir the mixture with a glass stirring rod and allow the mixture to digest for one hour. Add 50 mL of water and stir with the glass stirring rod. Allow the mixture to settle for 10 minutes. Carefully pour off the top half of liquid and discard any floating organic matter. Carefully add 20 mL of 3.0-*M* hydrochloric acid (pH 1.0). Stir the mixture with a glass stirring rod and allow the mixture to digest for one hour. Add 50 mL of water and stir with the glass stirring rod. Allow the mixture to settle for 10 minutes. Carefully pour off the top half of liquid and discard any floating organic matter. Filter the remaining soil and water through a funnel and filter paper into a 250-mL Erlenmeyer flask. After filtration is complete, place the remaining soil and filter paper in an oven at 105°C for 72 hours. Remove the samples from the oven and weigh the soil (weigh the filter paper and soil together, scrape the soil off the filter paper, and weigh the filter paper alone. Use these figures to determine the soil weight). Record your results in Figure 6-1.

3. Examine the role of organic matter in soil. Obtain six 4-in. plastic pots. Fill one pot with natural soil, one pot with 100 percent peat, and one pot with 100 percent sand. Place each of the pots on top of an inverted pot in a separate 5,000-mL beaker (see Figure 6-2). Water the pots to saturation with distilled water and collect 15 mL of the water that flows out of the drain holes in the bottom of each container in a labeled 20-mL collection vial. Place a drop of each drained water sample onto a glass microscope slide and examine under 1,000× magnification. Characterize the microorganism activity and record your observations in Figure 6-3.

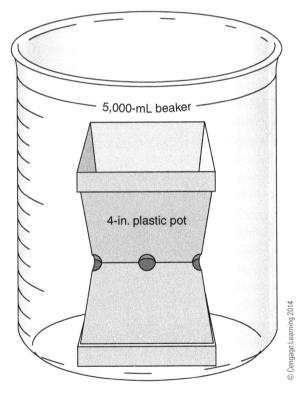

Figure 6-2
Pot drainage.

Sample	Initial Microbial Activity	Microbial Activity After Plant Growth
Natural soil		
100 percent peat		
100 percent sand		

Figure 6-3
Organic matter and microbial activity.

Let the pots stand for 48 hours. Water the pots as before until saturated. Again, collect 15 mL of the water that flows out of the drain holes in the bottom of each container in a labeled 20-mL collection vial. Using a conductivity meter, measure the soluble salt level in the drained water samples from each pot and record the result in Figure 6-4. Measure the pH of each drained water sample and record the result in Figure 6-4.

Drained Water Sample	Initial Soluble Salt Level	Initial pH	Soluble Salt Level After Plant Growth	pH After Plant Growth
Natural soil				
100 percent peat				
100 percent sand				

Figure 6-4
Organic matter and soluble salts.

Finally, germinate bean seeds in each pot and grow them in the lab under artificial lights for approximately two weeks. Water the seedlings with distilled water only. After the first true leaves have fully opened, perform the saturation and drainage collection process for a third time. Observe the microorganism activity in a drop of drained water sample from each pot on a glass microscope slide. Record your observations in Figure 6-3. Using a conductivity meter, measure the soluble salt level in the drained water samples from each pot and record the result in Figure 6-4. Measure the pH of each drained water sample and record the result in Figure 6-4.

CHALLENGE QUESTIONS

1. How do the organic matter determination techniques remove the organic matter from the samples?

2. How do the bean seedlings react to the different growing media? Why?

3. Which growing medium resulted in the highest soluble salt concentration? Why?

4. How did the different media alter the pH? Why?

CHAPTER 7

Soil Water

PURPOSE

The purpose of this chapter is to examine the interaction between soil and water.

OVERVIEW

Water and soil form an intricate and dynamic association. The soil's physical and chemical properties are profoundly affected by water present in the pores. The soil is the only source of water for most plants. Therefore, a comprehensive understanding of the soil–water interaction is crucial for soil managers.

Water that moves across the surface of soil is referred to as runoff and can be characterized as surface flow. This phenomenon has recently received extraordinary attention for its contributions to soil management and pollution mitigation. The Soil Erosion and Sediment Control Act (http://www.state.nj.us/agriculture/divisions/anr/agriassist/chapter251.html) mandates extensive efforts to manage surface water movement. These laws have been further extended to include certification programs for anyone making changes to the land through development or disturbance.

Water crosses the soil surface through a process known as infiltration. Water percolates through the soil profile. The rate of infiltration and percolation collectively characterizes the downward movement of water in the soil. Water is held in the soil by the forces of adhesion and cohesion within the soil pores. The size and shape of the pores control not only the water held but also the movement of water within the soil. Soil scientists measure these properties through a variety of laboratory procedures. However, water characteristics of soil can be mathematically estimated through knowledge of the particle size distribution or soil texture. This can be accomplished because the soil texture inherently describes the nature of the pore space between the particles. These mathematically predicted soil properties could be used to manage soil water.

ACTIVITIES

Soil water and its association with soil are either measured in the lab or mathematically estimated from the soil texture.

Equipment and Materials

10-in.-diameter, 6-in.-wide sheet metal ring

6-in.-diameter, 6-in.-wide sheet metal ring

watering can with water break nozzle

ruler

shovel

four 4-in. plastic pots

commercial potting media

bucket or pan

two sheets of plastic (15 in. × 15 in.)

Activity

1. *Determine the infiltration rate.* Use a double-ring infiltrometer to measure the infiltration rate of the soil. Select a location on the lawn that is level and dry. Be careful not to disturb the soil surface, leaving all vegetation in place. Press a 10-in.-diameter, 6-in.-wide sheet metal ring 1 to 2 in. into the soil (see the illustration in Figure 7-1). Next, press a 6-in.-diameter, 6-in.-wide sheet metal ring in the center of the 10-in. ring, also 1 to 2 in. into the soil. Fill both rings using a watering can with water break attached to a depth of 4 in. Measure the depth of water in the center ring every 5 minutes for the first 30 minutes and every 15 minutes thereafter until all the water has infiltrated the soil. Record the result on the graph found in Figure 7-2. Complete the same procedure again in a location with a bare soil surface and record your results in the graph found in Figure 7-2.

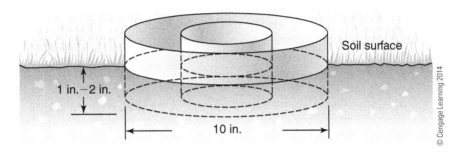

Figure 7-1
Double-ring infiltrometer.

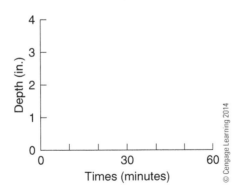

Figure 7-2
Infiltration rate.

2. *Measure the percolation rate.* Select another location on the soil surface that is level and dry. Dig three holes, at least 3 ft apart, in the form of a cylinder, taking care not to excessively disturb the surrounding soil and the soil in the bottom of the holes. Each hole should be 8 in. in diameter and 1 1/2 in. deep. Using a watering can with water break attached, slowly fill each hole with water. Measure the depth of water in the center of each hole every 5 minutes for the first 30 minutes and every 15 minutes thereafter until all the water has percolated into the soil. Record the result on the graph found in Figure 7-3.

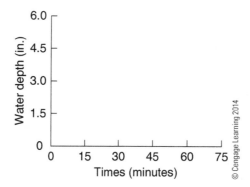

Figure 7-3
Percolation rate.

3. *Determine the water-holding capacity and drainable pore space.* Determine the water-holding capacity and drainable pore space of a natural soil and compare with artificial growing media. Obtain an air-dry natural soil sample and a commercial potting medium with the ingredients listed on the bag. Obtain four 4-in. plastic pots. Line two pots with plastic. Fill one of the plastic-lined pots with the natural soil sample and the other with the commercial potting medium to be tested. Firm the commercial potting medium as though you were planting into it. Fill each container with a measured volume of water until the media are completely saturated. Allow the media to sit for at least one hour. Add more water as necessary until the media are completely saturated and will not accept any more water. Record the total volume of water added in Figure 7-4. This represents the total pore space.

Next, place each of the pots on top of an inverted pot in a separate bucket or pan (refer to the illustration in Figure 6-2). This will capture the water that drains out of the media. Puncture the plastic lining in each pot through the drain holes. Be careful not to tilt or tip the containers while puncturing the plastic bags. Allow the water to drain for one hour. In Figure 7-4, record the volume of water drained from each container. Determine the proportion of drainable pore space by dividing the volume of water drained by the total volume of water added and multiply by 100. Enter your result in Figure 7-4. The percentage of drainable pore space should be between 20 percent and 30 percent for suitable commercial growing media. Determine the volume of available water held by deducting the water drained from the total volume of water added. Enter your result in Figure 7-4.

Quantity Measured	Natural Soil	Commercial Potting Media
Total volume of water added (mL)		
Water drained (mL)		
Proportion of water drained (%)		
Available water held (mL)		

Figure 7-4
Drainable pore space and available water held.

CHALLENGE QUESTIONS

1. What is the usefulness of these laboratory characterizations of water and its association with soil?

2. Why are the water-related measurements variable from soil to soil? Or, why do they differ between natural soil and artificial media?

3. How can the values in Figure 7-4 be extrapolated to apply to a one-acre growing area?

4. After completing the various analyses in this chapter, what other kinds of information would be required to develop a soil-water management program?

CHAPTER 8

Water Conservation

PURPOSE

The purpose of this chapter is to explore the various ways in which water can be conserved through soil management and other strategies designed to minimize supplemental irrigation requirements.

OVERVIEW

Water conservation through effective soil management will become increasingly important in the future. Soil managers can utilize their understanding of the soil–water interaction, in addition to the hydrologic cycle, to conserve water. In any conservation strategy, it is fundamental to determine a system's capabilities, its current state, and the rate of loss and to devise efforts for conservation. Soil's moisture-holding capacity can be estimated mathematically or directly determined through laboratory procedures. Measuring soil moisture levels can be accomplished through a variety of techniques. These range from the use of indicator plants to the use of neutron probes. Tensiometers also provide an effective measure of soil moisture levels, but many soil managers utilize evapotranspiration rates (ET_{rate}) to estimate the loss of water from the soil. Efforts to conserve soil water involve methods to improve the holding capacity of soil along with methods to minimize water loss. Water loss from the soil can occur in three ways. First, gravitational water percolates down through the soil profile. Second, water evaporates directly from the soil surface. Third, soil water moves through plants in a process known as transpiration. These three water losses can be quantified and monitored to conserve water.

Gravitational water is pulled down through the soil profile by the force of gravity. Water available to plants is held against the force of gravity between the soil particles in micropores. The higher the proportion of micropores contained within a soil, the higher its water-holding capacity. Additionally, organic matter in the soil can improve water-holding capacity. Therefore, amendments that increase the proportion of micropores or raise the organic matter content will often increase a soil's water-holding capacity. A water conservation program should consider opportunities to appropriately amend the soil.

Water available to plants is sometimes called *capillary water*. This is water held between field capacity and wilting point. Field capacity and wilting point are dynamic properties of the soil–plant system. The particular plant species and its interaction with the soil profoundly affect the field capacity and wilting point of the soil. A water conservation program capitalizes on this soil–plant interaction. Soil managers can use laboratory procedures to determine the range in which water is available to plants, or they can estimate the amount of available water in the field. The use of indicator plants and ET_{rate} can provide excellent information about the status of available water in a soil at any given time. These measures serve as a guide for how much and when to supplement soil moisture levels to achieve water conservation goals.

Finally, soil-water conservation is accomplished through careful plant selection and controlling plant water usage through various cultural practices designed to conserve water. Plant water consumption is based largely on its physiological stage of growth, nutritional state, water availability, and plant competition. All these can be manipulated to minimize plant water usage. Natural growth patterns in a plant's life cycle often serve to minimize water usage during times of low water availability. It is important to learn to recognize these growth patterns and incorporate them in an overall water conservation strategy.

ACTIVITIES

Water conservation can be achieved through the coordination of soil capabilities and plant management. This chapter provides an opportunity to isolate and observe each principle involved in a complete water conservation program.

Equipment and Materials

six 4-in. plastic pots
three buckets or pans
sand
silt
peat moss
plastic mixing tub
scale

Activity

1. *Estimate soil water-holding capacity.* Obtain sand, silt, and peat moss samples from the instructor. Fill one 4-in. plastic pot with sand. It may be necessary to cover the drain holes with small rocks to prevent the sand from pouring out, but be careful not to block the drain holes completely. In a separate container, mix 80 percent sand and 20 percent silt by volume. Fill one 4-in. plastic pot with the 80 percent sand and 20 percent silt mixture. Finally, mix 50 percent sand and 50 percent peat moss by volume. Fill one 4-in. plastic pot with the 50 percent sand and 50 percent peat moss mixture. Allow all three pots to air-dry for 72 hours inside the laboratory.

   Place each of the pots on top of an inverted pot in a separate bucket or pan (see the illustration in Figure 6-2). This will allow the water to drain out of the containers. Slowly pour 100 percent of the container volume of water into each container. Allow the containers to drain for 30 minutes. Carefully remove the soil-filled pots from the buckets and set aside (use these wet soil mixtures immediately in activity 2). Remove the inverted pots from the beaker and record the amount of water drained from each pot. Record this information in Figure 8-1. Deduct the amount of water drained from the water added to determine the amount of water held by each soil. Record this information in Figure 8-1.

2. Estimate the available water by weight. Weigh the three wet pots from activity 1. Record the weights in Figure 8-1. Allow the containers to air-dry in the laboratory. Weigh the pots again and record the weights in Figure 8-1.

Soil Mixture	Water Drained (mL)	Water Retained (mL)	Available Water by Weight (g)	Total Water Required by Bean Plants (mL)
Sand				
Sand/silt				
Sand/peat				

Figure 8-1
Soil water-holding capacity.

3. Observe the water conservation ability of different soil mixtures. Plant pregerminated bean seeds in the three pots used in activity 2. Keep an accurate record of the water applied to each plant for a period of three to four weeks. Water the plants on an individual basis and apply water only after the plants have begun to wilt. Compare the total amounts of water required by the bean plants growing in the different mixtures. Record your results in Figure 8-1.

4. Estimate the ET_{rate} of various plants. Most areas of the United States have access to daily ET_{rate} information. These rates are often available from the University Cooperative Extension Service or weather service within each state. For example, the University of Georgia posts daily ET_{rate} information on the Internet via the Georgia Automated Environmental Network at http://www.georgiaweather.net. The published ET_{rate} information can be used to schedule irrigation and manage soil moisture. The published ET_{rate} is usually adjusted for a particular plant or planting. Adjustments are made for the plant type, location, and cultural program by reducing the published ET_{rate} 25 percent to 75 percent. The adjustment is based on how fast the soil in a particular planting dries out relative to bare soil exposed to full sun. Additionally, the ET_{rate} can be aligned with actual soil moisture measurements to gain a very accurate record of the plant–soil–water relationship of a particular planting.

Record the published ET_{rate} for your area over a 10-day period in Figure 8-2. Adjust these values for some plantings in Figure 8-2. Use these estimates of available soil water to schedule

Date	Published ET_{rate}	Planting 1 ET_{rate}	Planting 2 ET_{rate}

Figure 8-2
ET_{rate} and watering plants.

the application of supplemental water to the selected plantings. Water might be applied when the moisture content of the soil is 50 percent of the total available water (capillary water) that can be held by the soil. To conserve water, only 60 percent of the soil water lost should be replaced.

CHALLENGE QUESTIONS

1. What types of information should be used when developing a plan to conserve soil water?

2. How can water be conserved through the use of soil amendments?

3. How does organic matter interact with water in the soil?

4. Why would replacing only a portion of the available water conserve soil water?

CHAPTER 9

Drainage and Irrigation

PURPOSE

The purposes of this chapter are to learn about removing water from the soil and to learn about the application of supplemental water to plants.

OVERVIEW

Grouping the topics of drainage and irrigation together is an ideal way to express the importance of both aspects of soil moisture management. The soil rhizosphere is a dynamic, living ecosystem. Like all other living systems, the rhizosphere requires constant influx and efflux of moisture. Therefore, the provision of drainage and irrigation cannot be overemphasized as the key ingredient to successful plant health management.

Soil and plant managers should utilize their understanding of soil physical properties to facilitate moisture influx and efflux in the rhizosphere. Infiltration rates can be manipulated with a wide variety of techniques to both conserve water and increase its movement into the soil profile. Percolation rates can be manipulated to maximize soil moisture available to plants but must also allow water to move through the profile, preventing oxygen depletion and relieving carbon dioxide buildup in the rhizosphere. Infiltration and percolation rates of soil supporting a naturalized group of plants are inherently optimized. Nature selects plants ideally adapted to the soil conditions, and the plants manipulate their soil environment to supply water, nutrients, protection, and anchorage. A soil manager can capitalize on this synergistic relationship between plants and the soil by establishing soil conditions that favor the desired crop.

Establishing desirable soil conditions begins by analyzing the existing soil conditions. Next, the differences between the desired conditions and the existing conditions are used to develop a plan for manipulating the soil either as preplant or as a part of the crop cultural program. Soil can be amended, mixed, aerated, or treated. Additionally, irrigation or drainage systems can be installed to further manage soil moisture for the crop. Economic considerations often influence the soil manager's recommendations for optimizing soil conditions depending on the potential value of the crop. Range management usually involves minimal soil manipulation, whereas sports turf can sometimes dictate tremendous soil managerial activities. A putting green may include complete soil modification along with an elaborate subsoil drain system. These drain systems might include pumps that can reverse the flow to cool the soil or provide suction to manage soil moisture levels. Artificial growing media used in container production of plants is another example of extreme soil manipulation to facilitate drainage and provide adequate moisture to plant roots.

ACTIVITIES

Drainage and irrigation can be used together to ensure desirable soil moisture levels. The following activities illustrate a few of the techniques used to more effectively manage soil moisture.

Equipment and Materials

marking paint	stopwatch
black marker	10-mL graduated cylinder
three wooden stakes (10 in. × 1 in. × 2 in.)	wooden spoon and plastic mixing bowl
core aerator	3 cups fine-textured soil
sand	three plastic tubes (1 1/4 in. diameter × 12 in. long)
peat moss	
double-ring infiltrometer	three pieces of cloth (2 1/2 in. × 2 1/2 in.)
watering can with water break nozzle	three rubber bands
dishwashing liquid	ten 4- to 6-oz cups

Activity

1. *Manipulate infiltration rate.* Select three 6 ft × 6 ft areas to perform infiltration tests. The areas must have the same soil and crop conditions and should be in close proximity (e.g., 10 ft apart in a triangular arrangement). Mark off each area with marking paint. Label each test plot with a wooden stake as A, B, or C. Plot A will serve as the control. Plot B should be aerated and top-dressed. Use a core aerator and pass over the area in two directions. Make sure the cores are 2 in. to 4 in. in length and leave them on the surface where they fall. Topdress the area with sand or an 80 percent sand, 20 percent peat moss mixture to a depth of 0.5 in. Be careful to apply the topdress uniformly over the area. Plot C should be left untreated like Plot A. The water being applied to Plot C will be modified.

   Install a double-ring infiltrometer to measure the infiltration rate of the soil near the center of each of the three locations you have selected (see Chapter 7 for infiltrometer details). Be careful not to disturb the soil surface inside the infiltrometer. Fill both rings using a watering can with water break attached to a depth of 2.5 in. Use ordinary tap water for Plot A and Plot B. Use tap water with 1 tablespoon of dishwashing liquid added for Plot C. Measure the depth of water in the center ring every 5 minutes for the first 30 minutes and every 15 minutes thereafter until all the water has infiltrated the soil. Plot your data on the graph found in Figure 9-1.

2. *Modify percolation rate.* Obtain 3 cups of a fine-textured natural soil sample. Place a homogenized (mixed rigorously in a bowl with a wooden spoon for three minutes) portion of the natural soil sample in a clear plastic tube 1.5 in. in diameter and 12 in. long mounted on a wire stand. Fill the tube to within 2 in. of the top, leaving this area open for the addition of water. Cover the bottom of the tube with a cloth and rubber band to prevent the soil from pouring out. See the illustration in Figure 9-2. Mix 50 percent coarse sand and 50 percent natural soil and fill a second clear plastic tube 1.5 in. in diameter and 12 in. long mounted on a wire stand to the same depth. Finally, mix 50 percent peat moss and 50 percent natural soil and fill the third tube. Label the three tubes.

   Examine the percolation rate of each sample, one at a time, by adding tap water to each tube (soil column), filling it to the top. Maintain a 3/4 to 1 in. head of water on top of the soil by adding water if necessary. Measure the depth of water percolation after 30 seconds, one minute, and every minute thereafter until the wetting front has reached the bottom of the column. It may be helpful to mark the column with a black marker at the end of each time period to track its progression through the soil. Complete this procedure for the other two soil columns. Plot the data you have collected on the graph found in Figure 9-3.

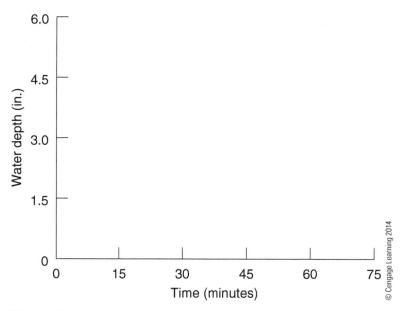

Figure 9-1
Infiltration rate.

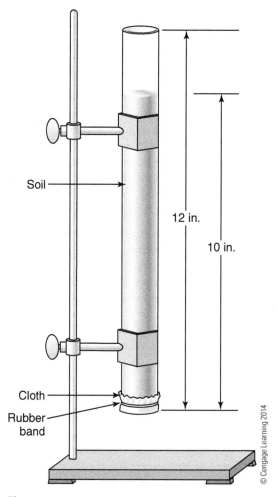

Figure 9-2
Soil percolation columns.

3. *Determine the uniformity of sprinkler application.* Select an area that is either irrigated with a sprinkler system or can be watered with a hose-end sprinkler. Mark off an area (minimum of 100 ft × 100 ft) that is to be covered by the sprinkler system. Place ten 4- to 6-oz cups at randomly selected locations within the area. Record the cup locations on a scaled drawing of the area. Run the sprinkler system for 30 minutes. Then measure and record the depth of water collected in each of the cups on the scaled drawing. Identify the three locations with the highest amount of water collected and the three locations with the lowest amount of water collected. If the collection of water varies substantially, determine ways that the system can be adjusted to improve the uniformity of application.

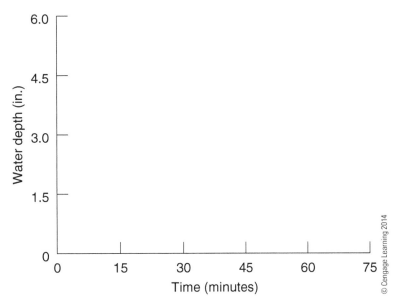

Figure 9-3
Percolation rate.

CHALLENGE QUESTIONS

1. Why does the infiltration rate change after aeration and topdressing? Why with dishwashing liquid?

2. How can the infiltration rate improve soil moisture available to plants?

3. What is the effect of the amendments on the percolation of water? Why?

4. How can a sprinkler system with poor uniformity cause pest problems for plants?

CHAPTER 10

Soil Fertility

PURPOSE

The purposes of this chapter are to learn how to recognize the nutritional needs of plants relative to the soil, and to develop a strategy for meeting those needs through soil management.

OVERVIEW

Providing adequate nutrition for plants is an exceedingly complex operation. Research has indicated that not only are the absolute levels of nutrients critical, but so are their levels relative to each other. This balancing of nutritional levels in the soil has been shown to dramatically impact plant resilience and health beyond what can be achieved with the traditional method of considering nutrient levels individually.

Soil chemistry and the soil reactions that characterize the exchange of essential nutrients will determine the availability of each element to uptake by plants. Soil managers must use their understanding of soil chemistry to develop a strategy to optimize nutritional levels within the rhizosphere. Similar to soil-water conservation, developing a soil fertility management plan begins with an analysis of a soil's capabilities. Next, through spectroscopy, the manager must determine the current level of the 14 essential nutrients derived from the soil. This is followed by a thorough investigation of the unique requirements of the particular crop being grown. Finally, a program can be developed to provide nutrition to the crop within the capabilities of the soil.

A soil's capabilities can be described through cation exchange capacity (CEC), pH, buffer capacity, base saturation, and water relations. All these soil properties combine to form the parameters for a plant nutrition program. It may be necessary to amend or modify a soil prior to planting in an effort to rectify problems associated with a soil's capabilities. Often, the incorporation of selected materials can increase CEC and buffer capacity, adjust pH, and improve water relations dramatically. Base saturation can also be modified through the judicious application of soil amendments. Once a crop is being grown, it becomes more difficult to manage these soil properties. Still, modifications can be made to existing crops that will improve a soil's fertility level and subsequent plant nutrition.

Soil samples should be sent to a testing lab for spectroscopic analysis to determine the present level of the 14 essential elements. When the results are available, the soil manager interprets them and formulates a strategy to reduce the availability of excessive nutrient concentrations and augment deficient nutrients (see Figure 10-1). This may result in the application of fertilizers, lime, sulfur, or other soil-fertility-modifying agents. It is important to recognize that the elemental concentrations available in the soil are not simply additive in nature. All of the reactions taking place must be considered when formulating the modification of soil fertility. The nutritional program must be customized to the crop being grown. The nutritional preferences of various plants differ significantly. In the past, there were broad, overall ranges for the essential elements required by plants. Today, these ranges are much more specific and vary from crop to crop. For example, cotton (*Gossypium* spp.) has a high requirement for zinc, whereas Bermudagrass (*Cynodon* spp.) requires very little. It is important to consider the crop that will be grown as well as the capabilities of the soil when formulating a nutritional program.

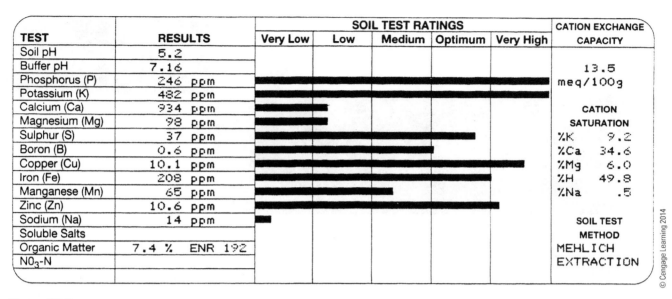

Figure 10-1
Soil sample lab report.

ACTIVITIES

Soil fertility is best analyzed with spectrographic techniques, like inductively coupled plasma (ICP) analysis. The interpretation of the test results and subsequent development of fertilization programs is a challenge. This chapter focuses on formulating fertilization strategies in relation to soil reports.

Equipment and Materials

rotary spreader

granular fertilizer

commercial fertilizer

Activity

1. *Determine the capabilities of a soil.* Review the soil test result in Figure 10-1. Formulate recommendations to resolve inadequacies in the CEC and base saturation levels. Detail your recommendations in Figure 10-2.

2. *Determine the fertility requirements of a crop.* Develop a fertilization program for a crop based on the soil sample report found in Figure 10-1. Search for information on corn nutritional requirements. Select a fertilizer from a local source. Make the selection based on the interaction between the nutrient levels indicated by the soil report and the nutrient sources in the fertilizer products. Indicate your fertilizer choice in Figure 10-2.

3. *Schedule a fertilization program.* Prepare a detailed plan for the application of fertilizer to 8,000 sq ft of Bermudagrass planting. Generally, Bermudagrass prefers 2–6 lbs of actual nitrogen per 1,000 sq ft during the growing season. This should be applied in increments of 1–2 lbs of actual nitrogen per 1,000 sq ft per application throughout the growing season. Select a fertilizer from a local source. It is appropriate to use a combination of several different fertilizer products if desired to meet the nutritional needs of a crop. Plan the nitrogen application schedule for the Bermudagrass plot. Include the amount of actual nitrogen and the amount of fertilizer that

Amendment	Amount	Purpose	Description
Topsoil		Add organic matter, raise CEC	
Lime		Add Ca and Mg, raise pH	
Others			
Corn crop fertilizer			

Figure 10-2
Recommendation for soil fertility management.

8,000-Square-Foot Bermudagrass Plot

Fert.	Jan.	Feb.	Mar.	Apr.	May	June	July	Aug.	Sept.	Oct.	Nov.	Dec.	Total
Lbs of actual N per 1,000 sq ft													

Figure 10-3
Fertility program.

should be distributed at each application in Figure 10-3. The total amount of nitrogen applied during the growing season is based on a variety of factors. The maintenance level, turfgrass usage, availability of irrigation, and soil properties are among the items that must be considered when selecting the appropriate amount of nitrogen to apply during a growing season.

4. *Apply granular fertilizer.* Select an area to apply fertilizer. Obtain a rotary spreader. Determine the product to be applied based on the fertilization program established for the area to be fertilized. Measure the area and determine the amount of fertilizer to be applied. Distribute the fertilizer uniformly over the area by calibrating the spreader to distribute 50 percent of the desired application rate and cover the area twice, passing in two directions. Record the application information in Figure 10-4.

```
Date _____
Weather _____
Technician _____
Size of area _____
Crop _____
Product applied _____
Lbs of actual N per 1,000 sq ft ____
Water applied _____
```

Figure 10-4
Fertilization record.

CHALLENGE QUESTIONS

1. What is the long-term strategy of soil fertility management?

2. Give three examples of nutrient interaction resulting in nutritional problems in plants (e.g., excessive phosphorus resulting in iron deficiencies).

3. How do slow-release nutrient sources change the determination of "actual" nutrient applications?

4. By law, what is meant by the term *guaranteed analysis*? How does this impact organic nutrient sources?

CHAPTER 11

Soil pH and Salinity

PURPOSE

The purpose of this chapter is to gain an understanding of the role of soil pH and salinity in plant nutrition.

OVERVIEW

Soil pH and salinity play a crucial role in the availability of nutrients to plants. What is referred to as soil pH and salinity is actually the pH or salinity of the soil solution. To measure these soil properties, a solution must be extracted from the soil and tested with carefully calibrated equipment. The procedure for extraction is important and cannot be neglected. Soil samples may be subjected to pure water extraction, buffer solution extraction, or both. Buffer solutions more accurately depict the reactions that take place within the biosphere and give an indication of the soil's resistance to change, or *buffer capacity*.

All plants are adapted to a particular pH range, some preferring a more narrow range than others. A soil manager's task is to recognize these preferred pH ranges and make adjustments where applicable. Changes to pH can be accomplished in a variety of ways. Lime is commonly used to raise pH. Sulfur, organic matter, or acidifying fertilizers are used to reduce pH. The amount of these materials that might be required is determined by the amount of change desired and the soil's resistance to change. Changes to pH should be made gradually over a period of time, usually six months to a year. Management of soil pH is the cornerstone of a soil fertility program.

Salinity of the soil solution is a reference to the quantity of ions dissolved in the soil water. Mineral nutrition of plants is accomplished by the absorption of these dissolved mineral ions from the soil solution. However, when the ion concentration becomes excessive, plants can be damaged and can encounter difficulty acquiring the nutrients they need. The most common treatment for excessive ion concentration is to thoroughly leach the soil to wash away or dilute the excessive ions. Some soils have a tendency to accumulate excessive ions and must be managed carefully to ensure that a crop's nutritional needs are met. Managing saline soils is one of the most challenging problems facing worldwide agricultural production.

ACTIVITIES

Soil pH and salinity are easily measured with lab equipment. The activities in this chapter allow you to become familiar with their operation. In addition, the growth response of plants can be observed under different pH and salinity regimes.

Equipment and Materials

natural soil sample

5-oz cup

10 mmol/L $CaCl_2$

distilled water

stirring rod

pH meter

conductivity meter

fifteen 4-in. plastic pots

peat-lite potting soil

20 marigold seeds

two 500-mL Erlenmeyer flasks

2.0-M hydrochloric acid

3.0-M sodium hydroxide

10 g table salt (NaCl)

20-20-20 liquid fertilizer

Activity

1. *Measure soil pH.* Obtain 2 tablespoons of a natural soil sample. Place the sample in a 5-oz cup. Add 2 oz of 10 mmol/L $CaCl_2$ solution and mix with a stirring rod for three minutes. Let the mixture stand for two hours. Stir the mixture again for one minute, filter through a paper towel, and measure the pH with a calibrated pH meter. Record your results in Figure 11-1.

2. *Measure soluble salts.* Obtain 2 tablespoons of a natural soil sample. Place the sample in a 5-oz cup. Saturate the sample with distilled water and mix with a stirring rod for three minutes to the consistency of a paste that glistens with water. Let the mixture stand for 24 hours. Extract the solution by suction filtration and measure the soluble salts with a calibrated conductivity meter. Record your results in Figure 11-1.

Sample	pH	Soluble Salts (dS/m)
1		
2		
3		

Figure 11-1
Soil pH and soluble salts.

3. *Effect on growth and development.* Germinate and grow 15 marigold plants. Each plant should be grown in a 4-in. plastic pot with a peat-lite growing mix. Fertilize the plants with 150 ppm of 20-20-20 liquid fertilizer at every watering. After the plants reach a height of 3 to 4 in., randomly select nine uniform plants for the following experiment. Label three of the plants solution A, three solution B, and three solution C. Water only with the solutions described next.

- Water three plants with the following:
  Solution A: Add 300 mL of distilled water to a 500-mL Erlenmeyer flask. Titrate with 2.0-M hydrochloric acid to pH 4.0. Titrate again with 3.0-M sodium hydroxide to pH 7.0.

- Water three plants with the following:
  Solution B: Add 10 g of table salt (NaCl) to 300 mL of distilled water in a 500-mL Erlenmeyer flask. Swirl the water until all of the salt is dissolved.

- Water three plants with the following:
  Solution C: Use distilled water only.

Water the plants as needed until observable differences exist between the treatments. Record daily observations in Figure 11-2.

Plant	Solution A	Solution B	Solution C
1			
2			
3			

Figure 11-2
Marigold plant development utilizing varying watering solutions.

CHALLENGE QUESTIONS

1. How can the pH of your soil sample be brought within the desired pH range for *Rhododendron* sp.? For hybrid tea roses?

2. What does the soluble salt concentration of a soil sample indicate about its desirability to produce crops?

3. Why did the marigold plants react the way they did to the various water treatments?

4. How does soil pH influence plant growth?

CHAPTER 12

Plant Nutrition

PURPOSE

The purposes of this chapter are to gain an understanding of the role of soil in providing essential elements to plants and to develop a nutritional program for a specific crop.

OVERVIEW

The acquisition of mineral nutrients requires a plant to expend a significant amount of energy. For example, the energy required to acquire and assimilate nitrogen can consume up to 30 percent of the total energy produced through photosynthesis. A soil manager should develop a plant nutrition program designed to minimize the energy required from a plant to obtain the 14 essential elements derived from the soil while considering physiological responses that may rely on nutritional deficits. It is sometimes counterproductive to overcome nutritional deficiencies when producing a crop. A key aspect of plant culture is to design management systems that maximize the efficient use of energy. The acquisition of essential nutrients from the soil is an integral part of such a management system.

The essentiality of a particular element in the growth and development of a plant is established by plant nutrition experiments. Plant scientists use three criteria to establish whether a nutrient should be considered "essential." First, the element must be present in order for a plant to complete its life cycle. Second, the element must be an integral part of a physiological process. Third, the element cannot be replaced by another element.

There are several examples of elements that meet only one or two of these criteria. Of course, an element is not granted the status of being essential by a selection committee. Instead, essentiality is achieved by an accumulation of a significant amount of data from experimental research and general acceptance by a majority of researchers in the field. Plants contain a vast array of other elements that are not considered essential (e.g., silicon [Si] and selenium [Se]). Currently, there are 17 elements that are considered to be essential. All are derived entirely from the soil except carbon (C), oxygen (O), and hydrogen (H). It is very important to recognize that the remaining 14 essential elements come only from the soil. Soil-derived elements are taken up by plants only in their ionic form dissolved in water. The ionic form of an element is derived from mineral salts, which dissolve in water trapped between soil particles. The porosity and ion exchange capacity of the soil determine its ability to provide nutritional elements to the trapped water. The pH (antilog of the hydrogen ion concentration) of the soil solution determines the tendency of mineral salts to dissolve, as well as the tendency for ions to be released from soil particles. This is why soil pH is important to plant nutrition. However, soil pH does not negate the importance of the porosity and ion exchange capacity of soils. More important, the entire process is based on soil water being taken up by plant roots.

To develop a fertility program that optimizes plant nutrition, the soil scientist must understand soil chemistry. In addition, the goals for the crop must also be determined. The goal of an agricultural crop is to maximize harvest, whereas the goal of an ornamental planting is to maintain a certain appearance. These are contrasting goals that require different nutritional programs. Often, the untimely application of a particular nutrient can negatively impact harvest or increase the incidence of pests. All these issues must be considered when planning a strategy for meeting the nutritional requirements of a crop.

ACTIVITIES

Plan a nutritional program to meet the nutritional goals of a specific crop by evaluating the soil and developing a fertilization program.

Equipment and Materials

nine 250-mL Erlenmeyer flasks

distilled water

1 tablespoon of clay soil

20-20-20 liquid fertilizer

fifty 4-in. terminal stem cuttings of *Plectranthus australis* (Swedish ivy)

Activity

1. *Evaluate a crop's nutritional requirements.* Contact a local extension agent or search the Internet to find the desirable range for phosphorus (P) in the soil to produce cotton. Compare this desirable range to the soil sample result in Figure 12-1. List your findings in Figure 12-2. Review the diagram in Figure 12-3 and formulate a strategy for meeting the phosphorus requirement of a cotton crop by completing Figure 12-2.

Element/Property	Lab Result	Element/Property	Lab Result
Phosphorus (ppm)	55	Manganese (ppm)	3.5
Potassium (ppm)	143	Iron (ppm)	10.0
Magnesium (ppm)	590	Copper (ppm)	0.4
Calcium (ppm)	88	CEC (mEq/100 g)	5.9
Soil pH	6.2		
Buffer pH	7.77	Base saturation: % K	5.8
Sulfur (ppm)	19	Base saturation: % Mg	12.4
Boron (ppm)	0.1	Base saturation: % Ca	50.0
Zinc (ppm)	3.6	Base saturation: % H	31.2

Figure 12-1
Soil sample report.

Crop	Desirable Range for Phosphrous	Phosphrous Level in the Soil	Soil Type
Phosphorus Management Strategy:			

Figure 12-2
Plant nutrition strategy.

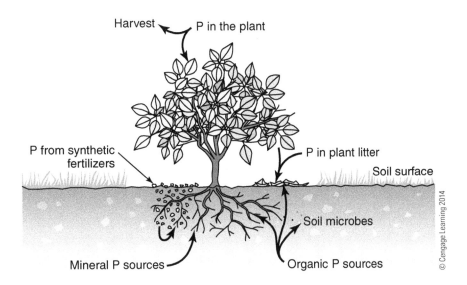

Figure 12-3
Phosphorus availability.

2. *Schedule the nitrogen (N) application for a corn crop.* Review the information in Figure 12-4 regarding the application of nitrogen to corn. Schedule the nitrogen application for a corn crop under the following conditions:

   - sweet corn
   - zone 7b in the southeastern United States
   - automatic irrigation available and unrestricted
   - sandy, unamended soil

   Complete the annual nitrogen application for the corn crop in Figure 12-5.

Corn Nitrogen Requirements

Expected Yield	Application	Nitrogen(lbs/acre)
Grain: 125 bushels/acre	Starter (2 in. 3 2 in.)[1]	35
	Sidedress[2]	106
Silage: 20 tons/acre	Starter (2 in. × 2 in.)	55
	Sidedress	165

[1] Starter fertilizer is applied 2 in. to the side and 2 in. below the seed.
[2] Sidedress fertilizer is applied in a band along the row.

Figure 12-4
Corn crop nitrogen recommendation.

Item	Jan.	Feb.	Mar.	Apr.	May	June	July	Aug.	Sept.	Oct.	Nov.	Dec.	Total
Lbs of actual nitrogen per acre													

Figure 12-5
Nitrogen application schedule for corn.

3. *Observe the effect of soil micelles as a micronutrient source.* Root fifty 4-in. terminal stem cuttings of *Plectranthus australis* (Swedish ivy) in 250-mL Erlenmeyer flasks. Prepare the culture solutions according to the following schedule:

- Solution A: 245 mL distilled water
  100 ppm 20-20-20 liquid fertilizer
- Solution B: 240 mL distilled water
  100 ppm 20-20-20 liquid fertilizer
  7 g clay loam soil
- Solution C: 245 mL distilled water

Label nine 250-mL Erlenmeyer flasks, three per solution. Place the flasks under artificial fluorescent lights, allowing 5 to 8 in. between the lights and the closest leaves.

Completely replace the solutions every five days by pouring out the old solution, rinsing the flask with distilled water, and refilling with fresh solution. Grow six cuttings in each solution (two cuttings per flask) until differences can be observed between the treatments. Record your observations in Figure 12-6.

Solution	Days in Culture	Apppearance of Foliage
A		
B		
C		

Figure 12-6
Nutrient solution culture.

CHALLENGE QUESTIONS

1. Why is a phosphorus source more effective when it is incorporated into the soil?

2. Why do crops like cotton, soybeans, and sorghum require high phosphorus?

3. How does the frequency of nitrogen application vary with soil texture? Why?

4. In the nutrient solution cultures, which nutrient solution provides the most complete nutrition to the plants? Why?

CHAPTER 13

Soil-Sampling and Testing

PURPOSE

The purpose of this chapter is to evaluate the sampling and testing techniques used to represent a soil environment.

OVERVIEW

A crucial element of soil-sampling and testing is the development of a strategy to collect a representative sample. Simply stated, the information desired must be obtained from the sample taken. The sampling procedure employed is a management decision. Several factors influence the steps taken to obtain a representative sample, including the intended soil use, depth, rhizosphere characteristics, amendment potential, disturbance, and viability. Once a sampling strategy has been devised, the testing can proceed.

If a site is under development, many soil cores may be taken to establish the structural integrity and variability of the site soil. Grading plans and structural footing designs depend on these soil characteristics. Further, septic systems and drainage patterns are also influenced by the underlying soil structure. If a crop is going to be grown, samples must be taken to determine water-holding capacities and nutritional conditions. The soil might also be tested for contaminants or internal drainage characteristics in an attempt to provide optimal rooting environments. Therefore, it is exceedingly important to establish why the sampling is being conducted before selecting the procedure for collecting the samples.

The focus of soil-sampling should be to obtain a soil-sample that represents the area to be utilized and to provide test results that can effectively represent a soil for the purpose of making managerial decisions. Whether these are nutritional, structural, or conservation oriented, the soil manager must rely on the test results from the samples taken. As a result, the soil scientist should carefully plan and execute a sampling procedure designed to meet the requirements of the site usage.

ACTIVITIES

Soil-sampling and testing plans are developed from information regarding the intended site usage. Students will develop soil-sampling and testing plans from information about the intended use of a site.

Equipment and Materials

none

Activity

1. *Develop a sampling and testing plan for land development.* Review the information regarding the site use plan for a residential development in Figure 13-1. Indicate the sampling locations and determine the laboratory tests to be performed. It is helpful to contact local geotechnical engineers and land planning companies for information about the types of soil tests performed to meet local ordinance requirements. Record this information in Figure 13-2.

2. *Develop a sampling and testing plan for a landscape planting.* Prepare a sampling and testing plan for the landscape site pictured and diagrammed in Figure 13-3. Record your plan in Figure 13-4.

3. *Develop a sampling and testing plan for an agronomic crop.* Prepare a sampling and testing plan for 180 acres of land to be planted with soybeans. The land is located in the Texas Panhandle. Record your plan in Figure 13-5.

Twelve Oaks Residential Subdivision

1. 50 acres of single-family units
2. 1/2-acre lots with 100-ft frontage
3. Municipal water
4. Septic systems
5. All electric
6. Concrete/curbed with 50-ft right-of-way
7. Utility easements back-of-curb
8. 2,500- to 3,500-sq-ft houses

Figure 13-1
Site planning information.

Site:_____

Address:_____

Use:_____

Zoning restrictions:

Soil-sampling location map:

Tests to be completed:

Figure 13-2
Soil-sampling and testing plan.

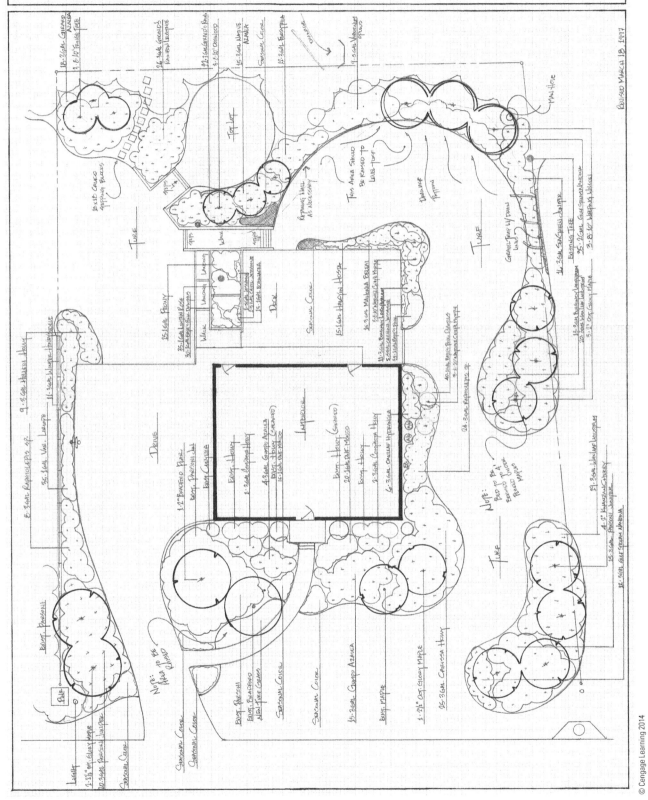

Figure 13-3

Residential site plan.

Site: _____	Zoning restrictions:
Address: _____	
Use: _____	
Soil-sampling location map:	Tests to be completed:

Figure 13-4
Soil-sampling and testing plan.

Site: _____	Zoning restrictions:
Address: _____	
Use: _____	
Soil-sampling location map:	Tests to be completed:

Figure 13-5

Soil-sampling and testing plan for soybean crop.

CHALLENGE QUESTIONS

1. What types of soil tests are required for development in the municipalities where you live?

2. How are soil test results representative of an area?

3. What are the limitations of sampling plans?

4. What are the costs associated with many of the soil-sampling plans you have developed?

CHAPTER 14

Fertilizers

PURPOSE

The purpose of this chapter is to explore various nutrient sources and combine them in a way that meets the nutritional requirements of plants while using caution to avoid excessive application.

OVERVIEW

The culture of plants often involves the augmentation of essential nutrients found to be deficient in the soil. A fertilizer is used as the source for the deficient nutrients. A careful review of the existing nutrient levels, their availability, and the chemistry involved with effectively supplementing deficient nutrients are the steps taken when selecting fertilizer products. A soil analysis reveals existing nutrient levels and must be completed before selecting fertility products. Nutrient levels are compared against those ranges recommended for the crop to be grown, and a strategy is developed for bringing each nutrient into the desirable range. Fertilizers are regulated by state laws that establish labeling requirements. The guaranteed analysis on these labels is usually the "minimum" amount of a nutrient present in the product. This value is less variable with synthetic nutrient sources when compared with organic nutrient sources. Therefore, the analysis can be misleading when utilizing organic products. Also, the term *actual* indicates the amount of a nutrient that will be immediately available to plants. Availability is controlled not only by the fertilizer's release mechanisms but also by soil properties and plant physiology. Some nutrient sources dissolve quickly in water and are immediately subject to rhizosphere dynamics. Other nutrient sources dissolve more slowly depending on temperature, water, and microbial activity. The method for determining nutrient availability or the "actual" amount of a nutrient in a fertilizer product must include all these considerations.

A guaranteed analysis describes the proportion of a nutrient included in a fertilizer product by weight. Therefore, it is a simple mathematical exercise to calculate the weight of a nutrient relative to the weight of the bulk fertilizer product. It is a much more challenging process to determine what proportion of a nutrient will ultimately show up in its ionic form in the soil solution surrounding plant roots. Because providing nutrients to a plant is the goal of a fertility program, the soil manager must strive to determine the proportion and timeline for nutrient availability in the rhizosphere. This should govern the selection of nutrient sources and fertility products.

The application method used to distribute fertilizer is also a consideration when selecting fertilizer products. Application equipment and its calibration are critical aspects of fertilizer product selection. In addition, the soil manager must determine the most effective and economical manner in which to deliver nutrition to the crop.

Granular fertilizers have become increasingly popular for their effectiveness and ease of application. Liquid fertilizers require careful equipment calibration, but can be equally effective. Uniformity is essential, and every effort should be made to distribute fertilizer products evenly over the entire area.

ACTIVITIES

Fertilizers are selected based on their nutrient sources and the requirements of a crop growing in a particular soil. The activities presented here are designed to familiarize you with the process of selecting fertilizers. In this lab, you select fertility sources in relation to physical and chemical properties of the soil. Determination of application rates and equipment calibration will be completed.

Equipment and Materials

rotary broadcast fertilizer spreader

dry, coarse sand or pelletized lime

marking paint

Activity

1. *Select nutrient sources.* Choose the most appropriate nutrient sources for a crop of trees and shrubs growing in the soil detailed in Figure 14-1. Complete your nutrient source selections and fertilizer blending recommendation by completing Figure 14-2.

2. *Determine the appropriate application rate.* Using the fertilizer product described in Figure 14-3, determine the application rate per 1,000 sq ft to supply 1 lb of nitrogen. Complete the remaining fertilizer calculations found in Figure 14-4.

3. *Calibrate application equipment.* Using a rotary broadcast spreader, determine the settings and application rates for coarse sand. On a blacktopped or concrete parking lot, mark off a square area 32 ft × 32 ft with marking paint. Weigh the amount of dry, coarse sand required to fill the hopper of a rotary spreader three-fourths full. Pelletized lime may be substituted for the coarse sand. Apply coarse sand or lime to the marked-off area using at least five different settings covering the range of settings available on the spreader. With each of the five applications, cover the area twice, as illustrated in Figure 14-5. First, pass back and forth, overlapping slightly on the edge of the distribution pattern. Second, make a similar application at right angles to the first coverage. Determine the amount of material dispensed after covering the area twice and record the results on the calibration graph in Figure 14-6. This can be accomplished by weighing the amount of material left in the spreader after covering the area twice at a particular setting and deducting from the original amount added. Be sure to refill the hopper three-fourths full with a known amount of material before repeating the procedure at each of the five settings. It is important to walk at the same pace throughout the application process.

Item	Value	Unit
Texture	Clay loam	N/A
pH	5.6	N/A
Buffer pH	7.25	N/A
CEC	14.3	mEq/100g
Organic matter	3	%
Nitrogen (N)	10	ppm
Phosphorus (P)	20	ppm
Potassium (K)	125	ppm
Calcium (Ca)	858	ppm
Magnesium (Mg)	95	ppm
Sulfur (S)	27	ppm
Boron (B)	0.3	ppm
Copper (Cu)	12	ppm
Iron (Fe)	104	ppm
Manganese (Mn)	48	ppm
ppm Zinc (Zn)	8	ppm
Sodium (Na)	16	ppm

Figure 14-1
Soil report.

Nutrient Source	Ionic Form(s)	Relative Salinity	Lbs Required per 50 lbs of Fertilizer

FIGURE 14-2
Nutrient source and blending recommendation.

SuperGreen Fertilizer

Guaranteed Analysis   27-4-12

Total Nitrogen 27%
    3.5% Ammoniacal Nitrogen
    23.5% Urea Nitrogen (12% slowly available poly-coated urea)
Available Phosphoric Acid (P_2O_5) 4%
Soluble Potash (K_2O) 12%
Iron (Fe)............................3%
Sulfur (S)............................8%

Primary Nutrient Sources: Poly-coated Urea, Urea, Diammonium Phosphate, and Muriate of Potash.

Secondary Nutrient Sources: Elemental Sulfur, Sulfur Oxides, and Sulfates of Iron.

FIGURE 14-3
Fertilizer label.

Amount of Nutrient Supplied per 1,000 Sq Ft	Amount of Fertilizer Required per 1,000 Sq Ft	Release Mechanism
1.0 lb of nitrogen		
0.12 lb of phosphorus		
0.5 lb of potassium		

FIGURE 14-4
Fertilizer calculations.

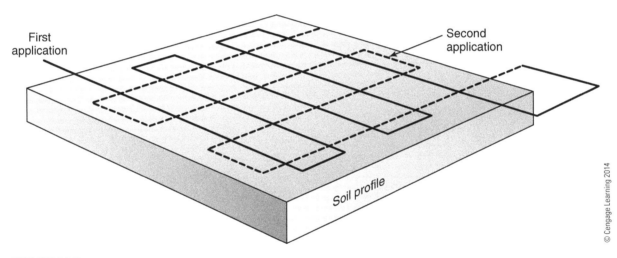

FIGURE 14-5
Granular fertilizer application pattern.

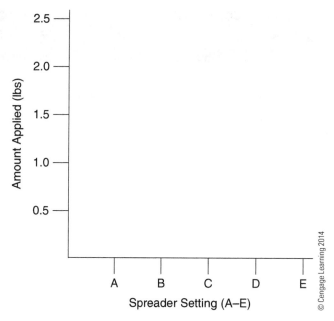

Figure 14-6
Calibration graph.

CHALLENGE QUESTIONS

1. What must be considered when determining the most economical nutrient source?

2. How might nutrient sources react when mixed together? Why are lime and fertilizer not blended together?

3. What nutrients are "actually" available to plants? When would this differ from what is actually in a fertilizer product?

4. What type of line best fits the calibration data developed in Figure 14-6? Why?

CHAPTER 15

Organic Amendments

PURPOSE

The purpose of this chapter is to evaluate the use of organic soil amendments and their contribution to soil–plant interaction.

OVERVIEW

The term *organic,* as it applies to soil amendments, has a very broad scope. Things that are considered to be organic amendments range from biosolids to compost. Biosolids or compost can consist of a wide variety of potential amendments. Animal manures, sludge, effluent, degraded plant material, animal by-products, and green manure all fall under the umbrella of *organic amendment.* These can be treated or untreated, incorporated or topdressed, and applied once or on a regular basis. Organic amendments might be used as nutrient sources, as physical property modifiers, or simply as a disposal mechanism. The purpose behind the application or incorporation of these amendments determines the suitability of their use. For example, if the reason for using an organic amendment is to reduce volume in a waste disposal stream, then the influence on the soil–plant environment is a secondary consideration. Evaluation of the use of organic amendments often includes factors that reach far beyond the impact on the soil.

The use of organic amendments as a nutritional source can be very effective. The process would begin by obtaining a thorough analysis of the soil and the proposed amendment. These results would then be compared to the crop's nutritional requirements and a review of the soil reactions that could be reasonably predicted. Of course, organic nutrient sources are exceedingly complex, and their reaction in the soil is difficult to predict. However, some margin of confidence can be achieved by considering the composition of the proposed amendment and previous experience with its use. A new or untested organic nutrient source should be tested on a small scale before its incorporation into the established soil fertility program.

The use of organic amendments as a modifying agent for a soil's physical characteristics can also be effective. Analyses from the soil and the proposed amendment should be compared. Organic materials often act to adhere soil particles into aggregates, which can improve the ratio of micropores to macropores in the soil, thereby improving both aeration and water-holding capacities. These modifications are variable in their duration and may be excessively temporary to be warranted. Determining the appropriate amount of amendment to add and the depth at which to incorporate it must be considered carefully. The rhizospheric environment is influenced not only by its composition, but also by the composition of the soil profile below it. Water and air must be drawn through the rhizosphere, and this is largely determined by the physical characteristics of the underlying soil profile. Finally, the quantity of organic amendment must be sufficient to impart the desirable effects. Some experts suggest that organic amendments below 25 percent by volume have only minimal effects. Often, the quantity required and depth of incorporation are grossly underestimated.

ACTIVITIES

Soil organic amendments are utilized as nutrient sources, physical property modifiers, or both. These effects can be observed in the lab by analyzing various combinations of soil and organic amendments before their application in the field.

Equipment and Materials

- 1 gal natural soil sample
- 1 gal compost
- 1 gal milled pine bark
- 250 mL 10 mmol/L $CaCl_2$
- calibrated pH meter
- conductivity meter
- glass stirring rod
- scale
- measuring cup
- three 6-oz cups
- oven
- paper bags
- three 4-in. plastic pots

Activity

1. *Measure a chemical effect of organic amendments.* Obtain 1 cup of a natural soil sample. Measure the natural soil's pH by placing 2 tablespoons of the sample in a 6-oz cup. Add 3 oz of 10 mmol/L $CaCl_2$ solution and mix with a glass stirring rod for three minutes. Let the mixture stand for two hours. Stir the mixture again for one minute; filter and measure the pH with a calibrated pH meter. Record your results in Figure 15-1. Measure the natural soil's soluble salts by placing 2 tablespoons in a 6-oz cup. Saturate the sample with distilled water and mix with a glass stirring rod for three minutes to the consistency of a paste that glistens with water. Let the mixture stand for 24 hours. Extract the solution by suction filtration and measure the soluble salts with a calibrated conductivity meter. Record your results in Figure 15-1.

   Complete the above two procedures for organic compost material either obtained commercially or composted. Obtain 1 cup of compost. Measure the compost's pH by placing 2 tablespoons of the sample in a 6-oz cup. Add 2 oz of 10 mmol/L $CaCl_2$ solution and mix with a glass stirring rod

Property	Natural Soil	Compost	Mixture
pH			
Soluble salts (dS/m)			

Figure 15-1
Chemical properties of soil, amendment, and amended soil.

for three minutes. Let the mixture stand for two hours. Stir the mixture again for one minute; filter and measure the pH with a calibrated pH meter. Record your results in Figure 15-1. Measure the compost's soluble salts by placing 2 tablespoons in a 6-oz cup. Saturate the sample with distilled water and mix with a glass stirring rod for three minutes to the consistency of a paste that glistens with water. Let the mixture stand for 24 hours. Extract the solution by suction filtration and measure the soluble salts with a calibrated conductivity meter. Record your results in Figure 15-1.

Finally, measure the pH and soluble salts of the natural soil amended with compost. Thoroughly mix 1/2 cup of natural soil with 1/2 cup of compost. Measure the mixture's pH by placing 2 tablespoons in a 6-oz cup. Add 3 oz of 10 mmol/L $CaCl_2$ solution and mix with a glass stirring rod for three minutes. Let the mixture stand for two hours. Stir the mixture again for one minute; filter and measure the pH with a calibrated pH meter. Record your results in Figure 15-1. Measure the mixture's soluble salts by placing 2 tablespoons in a 6-oz cup. Saturate the sample with distilled water and mix with a glass stirring rod for three minutes to the consistency of a paste that glistens with water. Let the mixture stand for 24 hours. Extract the solution by suction filtration and measure the soluble salts with a calibrated conductivity meter. Record your results in Figure 15-1.

2. *Measure physical effects of organic amendments.* Estimate the water-holding capacity of natural soil, soil–compost mixture, and soil–pine bark. Fill one labeled 4-in. plastic pot with natural soil. In a separate container, mix 70 percent natural soil and 30 percent compost by volume. Fill one labeled 4-in. plastic pot with the mixture. Finally, mix 70 percent natural soil and 30 percent milled pine bark by volume. Fill one labeled 4-in. plastic pot with the mixture. Empty each plastic pot into a labeled paper bag and oven-dry the soil samples at 105°C for 72 hours. Remove the samples from the oven and weigh them. Record the sample and empty pot weights in Figure 15-2. Measure the volume of soil or soil mixtures in the measuring cup. After recording the weight and volume of each sample, refill each of the plastic pots with its

Property	Natural Soil	Soil–Compost Mixture	Soil–Pine Bark Mixture
Empty pot weight (g)			
Soil dry weight (g)			
Saturated weight (g) (deduct pot weight)			
Dry volume (g)			
Bulk density (g/cm^3)			
Water-holding capacity (mL)			
Porosity (% by volume)			

FIGURE 15-2

Physical properties of natural soil, soil–compost mixture, and soil–pine bark mixture.

respective soil or soil mixture. Carefully saturate the soil or soil mixtures with tap water, allowing water to flow freely out of the drain holes. Let them stand for two hours and saturate the samples again. Finally, weigh the saturated samples and pots together. Complete the information in Figure 15-2.

CHALLENGE QUESTIONS

1. How would you explain the variations in chemical properties between soil and organically amended soils?

2. How would organic amendments affect the availability of nutrients to plants?

3. What influence do organic amendments have on the physical properties of soil?

4. What types of inoculants and plant extracts are used as organic amendments?

CHAPTER 16

Tillage and Cropping Systems

PURPOSE

This chapter examines the effects of various processes utilized by growers to manage soils for the purpose of growing plants.

OVERVIEW

Soils are frequently modified or manipulated to improve their productivity. Such practices can improve or damage the soil, especially when used repeatedly over a period of time. Often, mixing and loosening of soil will improve aeration and modify the soil's structure to create more favorable conditions for root growth. Rototilling, discing, harrowing, and plowing are all examples of this principle. However, applied incorrectly these cultural practices can be damaging to the soil. Some experts suggest that the benefit of adding amendments into the soil is actually more from the process of their incorporation than from the amendments themselves. Repeated soil modifications can damage the soil structure and, ultimately, reduce a soil's productivity. The soil manager must be aware of the immediate and long-term effects of tillage and cropping systems. Recently, intercropping systems have experienced a resurgence of popularity. The effectiveness of these relative to soil management has yet to be established.

Any method of turning or mixing the soil disturbs the rhizosphere to some degree. Plant roots are cut, microbial flora are homogenized, and soil aggregates are formed and broken. Fresh air is incorporated, moisture is lost, and seeds may be buried or brought to the surface. All these changes may be summarily beneficial for a particular crop, but there are detrimental effects that must also be considered. Turned or mixed soil is much more susceptible to compaction, erosion, leaching, loss of microbial populations, and solarization. The threat of erosion has prompted many soil managers to devise cropping systems that eliminate or minimize tilling of the soil. These "no-till" cropping systems rely on well-aerated soil and carefully applied weed control, but can improve water conservation and prevent erosion. These soil management challenges apply to agronomic as well as horticultural growing situations. A soil manager creates the most appropriate blend of tillage and cropping techniques to maximize production, while at the same time conserving soil and water.

ACTIVITIES

Soil management includes tilling and cropping systems uniquely tailored to individual growing situations. These activities are designed to familiarize you with the effects of soil cultivation practices.

Equipment and Materials

soil probe

rototiller

sprinkler

grass seed

nonselective, biodegradable herbicide

scale

hand rake

wheat straw

three clear plastic tubs (1 ft deep × 1 ft × 20 in.)

natural soil

pine straw

lava rock

2-gal watering can with water break

black marker

Activity

1. *Measuring the effect of rototilling.* Mark off two adjacent 32 ft × 32 ft square, level areas on the campus. Spray out all the vegetation with a nonselective herbicide. After the vegetation has died, rototill one of the areas in two directions to a depth of 8 in. Leave the other area untilled. Lightly rake the tilled area smooth with a hand rake. Obtain 6-in. soil cores from four different locations within each of the marked-off areas. Average the bulk densities of the four soil cores collected from each area by recording their volume and weight in Figure 16-1. Next, mathematically estimate the porosity and record the result in Figure 16-1.

2. *Growth rate variance.* Measure the growth rates of seedlings planted in the areas prepared for activity 1. Distribute grass seed (e.g., tall fescue) at the recommended rate over both areas. Mulch the areas with wheat straw to a depth of 2 to 3 in. Water both areas lightly and evenly as necessary to keep the soil surface moist until the seed has germinated. Continue to water both areas as necessary to maintain adequate soil moisture for grass plant development. Estimate the grass growth every day by placing a sheet of paper on a uniform, representative stand of grass blades and measuring its height until the grass grows to 3 in. Record your observations on the graph in Figure 16-2.

3. *Observe the effects of soil coverings.* Fill three clear, plastic tubs (1 ft deep × 1 ft × 20 in.) with natural soil to a depth of 8 in. Make sure the tubs have six 1-in.-diameter holes in the bottom to allow water to drain out. Mark this depth on the outside of the container with a black marker. Lightly level the top of the soil, being careful not to compact the soil. Cover the soil in one of the tubs with 2 in. of pine straw or bark mulch. Cover the soil in another tub with 2 in. of lava rock. Leave the soil in the third tub uncovered.

   Water the three tubs using a 2-gal watering can with water break attached. Carefully distribute 2 gal of water evenly across the top of the soil contained in each tub, allowing the water droplets to fall 6 to 8 in. before entering the tubs. Allow the water to completely infiltrate the soil. Observe the infiltration and percolation of water in the three tubs. After the water has completely infiltrated the soil, mark the soil level with a black marker. Measure the change in the soil level or settling and record the result in Figure 16-3.

Area	Volume (cm3)	Weight (g)	Bulk Density (g/cm³)	Porosity (%)
Undisturbed 1				
Undisturbed 2				
Undisturbed 3				
Undisturbed 4				
Undisturbed average				
Tilled 1				
Tilled 2				
Tilled 3				
Tilled 4				
Tilled average				

Figure 16-1
Rototilling effects on bulk density and porosity.

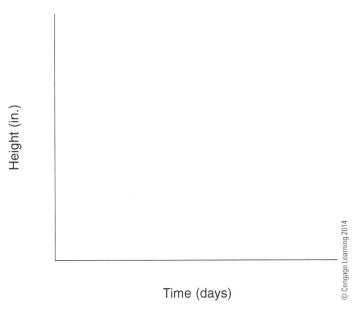

Figure 16-2
Growth rate.

Soil Surface	Settling (in.)	Observations (infiltration rate, wetting front progression, distribution, surface, etc.)
Bare soil		
Pine straw		
Lava rock		

Figure 16-3
Mulch effects.

CHALLENGE QUESTIONS

1. How long do the effects of rototilling last?

2. What differences occurred in the development of the grass in activity 2? Why?

3. How can the benefits of cultivation activities be applied to existing plantings?

4. How do mulching materials buffer the soil? How does nature buffer the soil from the impact of rain?

CHAPTER 17

Horticultural Uses of Soil

PURPOSE

The purpose of this chapter is to explore the various uses of soil and soil science in horticultural applications.

OVERVIEW

The horticulture industry is widely varied. It ranges from fruit and vegetable production to ornamental and turf management to production of plants in containers. The use of soil science in these areas is equally varied. However, regardless of the application, the basic principles of soil science remain the same. A soil's physical and chemical properties form the center of any soil application. Both the horticulturist and the soil scientist must utilize their understanding of these most basic soil properties to manage the soil for any growing application.

Some horticultural applications can be particularly demanding of the soil. Often, an ornamental planting exists on very poor-quality soil. In other instances, fruit or vegetable production is heavily reliant on good soil management. In every growing situation, representative soil samples must be collected and analyzed to establish a baseline for soil management programming. Physical and chemical analyses reveal the existing soil conditions. In addition, pest populations in the soil must be characterized before unwittingly embarking on susceptible crop production. Vegetable and fruit crops are sensitive to pest pressures existing in the soil. Some knowledge of the crop to be grown and its cultural requirements leads to a soil management plan that can maximize crop performance.

Sometimes, the requirements of a specific growing system exceed the soil's capacity. For example, the use of artificial growing media in container plant production facilitates drainage, maximizes water-holding capacity, and allows for nutritional control, all of which would be unavailable through the use of natural soils in containers. The physical and chemical properties are applied to artificial media components just as they are to natural soils. The plant–soil relationship remains the same, even when the soil is an artificial medium.

Whether the soil environment is a short-term application, as with annual or containerized crops, or a long-term application, as with fruit orchards or ornamental plantings, the requirement for soil management still exists. The use of soil science to provide moisture, nutrition, anchorage, and protection for plant roots is essential.

ACTIVITIES

The use of soils for horticultural applications involves a wide variety of growing situations. Soil scientists must apply the principles of soil science to each of these situations to maximize the potential of the soil and the crop. Growing fruits, vegetables, and ornamental plantings requires these horticultural applications.

Equipment and Materials

two natural soil samples (1 lb each) with associated lab reports

nine containerized (4-in.) annual plants grown in artificial soil media

nine plastic growing containers (10-in. diameter × 10-in. depth)—

or excavate a planting area where sections can be constructed with different soils separated by dividers.

1 ft^3 organic matter

30 lbs clay loam soil

30 lbs loam soil

30 lbs sandy loam soil

spade

labels

Activity

1. *Recommend modifications to existing soil.* Consider the crop requirements detailed for the production of peaches listed in Figure 17-1. Compare this information to the soil sample provided by your instructor. Make recommendations for soil management and modification in Figure 17-2. Be sure to include both physical and chemical modifications, in addition to soil-borne pest management strategies.

2. *Develop a soil management program.* On the basis of the tomato production requirements given in Figure 17-3, devise a soil management program for the soil sample provided by your instructor. Record your recommendations in Figure 17-3.

3. *Observe the movement of plant roots through transitioning soil textures.* Set up a growing environment with conflicting soil textures. Obtain nine containerized (4-in.) annual plants grown in artificial soil media. Prepare three different soil environments: one prepared by incorporating 40 percent by volume organic matter into sandy loam soil, another consisting of loam soil, and the third of clay loam soil. Fill nine plastic growing containers (10-in. diameter × 10-in. depth), three of each soil type. Or, establish planting areas with the different soil types (this will avoid the problems associated with natural soils, drainage, and containerized growing conditions). Plant and label each of the annual plants in one of the soil types. Water the plants with only enough water to settle the soil around the root ball but not saturate the soil. Place the plants under artificial growing lamps or in a greenhouse. Maintain day and night temperatures of 25°C and 13°C, respectively. Water the plants as necessary with a maximum of 20 oz per application to avoid saturating the surrounding soil. Grow the plants for three to four weeks.

    Carefully remove one-half of the root ball from each container. This should be accomplished by inserting a spade into the center of the container slightly to one side of the plant crown. Tilt the spade to the side away from the plant and pull the half root ball out of the pot, as illustrated in Figure 17-4. Examine the root development out from the original root ball into the surrounding soil. Record your observations in Figure 17-5.

Item	Requirement	Recommendation
Soil texture	Sandy loam	
pH	6.0 to 6.5	
N	Nitrogen applications are based on the previous season's growth. Nitrogen application should not exceed 20 oz of N per 1 in. of trunk diameter measured 1 ft above ground level.	
P	40 to 65 ppm	
K	90 to 220 ppm	
Ca	1,200 to 1,400 ppm	
Mg	400 to 600 ppm	
S	30 to 50 ppm	
Fe	120 to 170 ppm	
Mn	20 to 45 ppm	
Cu	2 to 5 ppm	
B	0.5 to 0.8 ppm	
Zn	2 to 4 ppm	
Moisture	Uniform availability	
Organic matter	5% to 10%	
Preparation	Loosen to a depth of 18 in.	
Other	Nematode populations must be identified and treated prior to planting.	

Figure 17-1

Recommendations for peach production in the southeastern United States.

Characteristic	Existing	Desirable	Modification Recommendation
Texture			
Structure			
pH and buffer			
CEC			
Nematodes			
Fungi			
Bacteria			
P			
K			
Ca			
Mg			
S			
Fe			
Zn			
Na			
Organic matter			
Depth			

Figure 17-2

Soil management and modification program for peach production in the southeastern United States.

Item	Requirement	Recommendation
Soil texture	Sandy loam to clay loam	
pH	6.0 to 6.5	
N	5 to 25 ppm	
P	50 to 60 ppm	
K	75 to 150 ppm	
Ca	975 to 1,250 ppm	
Mg	400 to 500 ppm	
S	20 to 30 ppm	
Fe	100 to 150 ppm	
Mn	15 to 40 ppm	
Cu	3 to 6 ppm	
B	0.5 to 0.8 ppm	
Zn	2 to 4 ppm	
Moisture	Uniform availability	
Organic matter	5% to 10%	
Preparation	Loosen to a depth of 18 in.	
Other		

Figure 17-3

Tomato production soil requirements and recommendations.

Figure 17-4
Removal of one-half of the root ball with a spade.

Plant	Soil	Root Growth and Development
1	Amended	
2	Amended	
3	Amended	
4	Loam	
5	Loam	
6	Loam	
7	Clay loam	
8	Clay loam	
9	Clay loam	

Figure 17-5
Observations of root growth and development after transplanting from the original root ball.

CHALLENGE QUESTIONS

1. How would you determine which soil modifications or management practices would be economically feasible?

2. What options exist for rectifying calcium deficiency in tomato plants?

3. What problems are often associated with establishing container-produced plants into a natural soil?

4. How can plant tissue samples be used to manage plant nutrition for horticultural crops?

CHAPTER 18

Soil Conservation

PURPOSE

The purposes of this chapter are to learn the importance of soil conservation and to practice some techniques currently used to conserve soil.

OVERVIEW

The importance of soil was studied in Chapter 1. By recognizing soil's importance and coupling that with an understanding of the magnitude and mechanisms of soil loss, the value of soil conservation becomes apparent. This was vividly illustrated by Hugh Hammond Bennett in early April 1935 as he testified before the Senate Public Lands Committee. While Bennett built a case for a permanent governmental agency that would focus on soil conservation, the skies were darkened by dust-filled winds blowing in from the central U.S. plains. The committee paused for a moment to look out the windows, and subsequently created the Soil Conservation Service (SCS). The SCS was the predecessor for what currently is called the Natural Resources Conservation Service (NRCS). Visit the NRCS Web site at http://www.nrcs.usda.gov/ to learn more about this service. The conservation of soil today is no less important, but much less popular. Many new laws and regulations have been enacted by both federal and state authorities to manage soil conservation. An example of these can be found at http://www.gaepd.org/Documents/rules_exist.html. Some states require training and certification for individuals involved in soil manipulations.

Soil scientists must include in every soil management plan specific strategies to not only retain the soil in a given location but also keep the soil productive by preventing its degradation. Strategies include cropping systems that minimize soil cultivation and soil modification requirements. Soil-building crops such as green manures, leguminous crops, and genetically modified organisms (GMOs) can all be used to conserve soil. World food production is reliant on continuously increasing yields, which are, in turn dependent on effective soil conservation practices. Soil management and food production require effective soil conservation.

ACTIVITIES

Soil conservation involves recognition of the magnitude and character of soil losses and the implementation of strategies to minimize these losses. The activities in this chapter are designed to illustrate both points.

Equipment and Materials

20 in. × 20 in. × 3/4 in. plywood

80 lbs natural soil sample

hose-end sprinkler or automatic sprinkler system

materials determined by the erosion control plan in activity 2

Activity

1. *Determine the amount of soil loss to water erosion.* Place a 20 in. × 20 in. × 3/4 in. piece of plywood on a level area. Mound 80 lbs of dry natural soil in a pile on the center of the piece of plywood. Set up an overhead sprinkler capable of applying 1 in. of water per hour onto the mound of soil. Run the sprinkler for two hours each day for four days. Make daily observations of the condition of the soil mound and record these observations in Figure 18-1. After the four days of overhead irrigation, collect the soil remaining on the plywood, dry, and weigh. Record the weight in Figure 18-1.

2. *Prevent soil erosion.* Locate an area where soil is being eroded by wind or water. Devise a plan to eliminate the erosion. The plan could include detention or retention ponds, filter strips, ground covers, incorporation of polymers, or windbreaks. Install the plan and monitor its effectiveness. Record your plans and observations in Figure 18-2.

3. *Compare predicted erosion to actual erosion.* Using the Universal Soil Loss Equation (USLE), estimate the erosion that would occur from the soil set up for activity 1. Enter the estimate in Figure 18-3. Extrapolate the actual erosion that took place to the same units used in the USLE and enter the result in Figure 18-3. As an alternative, use computer-based erosion-estimating tools found on the Internet.

Natural Soil Weight	Eroded Soil Weight	Observations of Erosion
80 lbs		

Figure 18-1
Water erosion.

Drawing of the erosion	
Erosion prevention plan	
Effectiveness of the plan	

Figure 18-2
Soil erosion prevention.

Equation	Result
USLE (A = R*K*LS*C*P)	
Actual erosion in activity 1	
Computer-based estimate	

Figure 18-3
Soil erosion prediction.

CHALLENGE QUESTIONS

1. How does the precipitation rate utilized for activity 1 compare to precipitation rates for rainfall in your area?

2. What techniques can be used to minimize the erosion occurring in activity 1?

3. What large-scale erosion control measures are required by law for land development in your area?

4. Why do the mathematical predictions differ from the actual erosion as measured by activity 1?

CHAPTER 19

Urban Soil

PURPOSE

The purposes of this chapter are to analyze the variability of urban soils and to devise soil management strategies for their use.

OVERVIEW

The soils found in urban areas are under increasing demand to provide not only structural support but also ecological maintenance, plant nutrition, and recreation facilities. These requirements are a challenge for urban soils, which are typically disturbed, modified, or contaminated. Soil scientists must extend their knowledge of soil management to incorporate the special situations presented by urban soils. As with other soil management planning activities, representative samples are collected and analyzed to establish the existing conditions. However, urban soils may be very difficult to sample in a representative fashion. Small areas may be drastically different from other immediately adjacent small areas. These conditions are often the result of significant soil moving and mixing activities during the development process. In addition, the incorporation of soil contaminants may be the result of development activities. These development processes challenge traditional soil-sampling techniques and the practice of formulating soil management plans based on a blend of multiple samples. A soil contaminant may be localized, whereas a blended sample would suggest more broad-based contamination. The sampling and testing strategy for urban sites must be carefully designed to account for the potential of significant variation in soil composition and soil uses.

Environmental impact is increasingly scrutinized, and soil offers an array of environmental mediation opportunities. Leadership in Energy and Environmental Design (LEED; http://www.usgbc.org/DisplayPage.aspx?CMSPageID=1988) certification programs rely on soil engineering techniques to mitigate the environmental impact of construction projects. While such strategies have traditionally been cost-prohibitive, a new wave of environmental concern is creating opportunity for soil management plans only dreamt about by soil scientists in the past.

ACTIVITIES

Urban soils vary significantly, and their uses can be quite demanding. The activities of this chapter explore ways to sample urban soils and devise management plans for their use.

Equipment and Materials

urban site
10-mm-grid vellum drafting paper
    (18 in. × 24 in.)
    or computer-aided drafting
    program
straightedge

soil lab testing equipment
urban soil sample

woodland soil sample
agricultural soil sample

Activity

1. *Develop a soil map for an urban plot of soil.* Prepare a scaled drawing of an urban site. Use 10-mm-grid vellum to prepare the drawing or utilize a computer-aided drafting program. Use contour lines or spot elevations to depict significant changes in elevation. Detail the use and purpose of each area included on the plan. Select sampling sites and number their location on the plan. Obtain the samples and analyze them. Complete the soil survey in Figure 19-1.

Sample Location	Texture	pH	Infiltration Rate	Profile Characteristics	Organic Matter	Contaminants
1						
2						
3						
4						
5						
6						
7						
8						
9						
10						
11						
12						

Figure 19-1
Urban soil survey.

2. *Prepare a soil management plan.* Using the soil survey from activity 1, complete a soil management plan. Include initial soil modifications, cost estimates, and maintenance schedules in your plan. Highlight your plan in Figure 19-2.

3. *Urban soil degradation.* Compare the soil samples taken from an urban site, a woodland site, and an agricultural site. Compare the soil samples from the three sites with regard to their physical characteristics. Record the comparison in Figure 19-3.

Activity	Location	Cost Estimate
Tilling		
Amendments		
Aeration		
Drainage systems		

Figure 19-2
Urban soil management plan.

Chapter 19

Property	Urban Soil	Woodland Soil	Agricultural Soil
Texture			
Structure			
Profile characteristics			
Infiltration rate			
Color			
Uniformity			
Water-holding capacity			
Porosity			

Figure 19-3
Soil comparisons.

CHALLENGE QUESTIONS

1. What type of soil dominates urban sites in your area?

2. How are soils and water runoff managed by the municipal planning departments?

3. How can you resolve the inherent conflict between the structural requirements and plant requirements of soil on an individual urban site?

4. What types of contamination are common in the urban soils in your area?

CHAPTER 20

Government Agencies and Programs

PURPOSE

The purpose of this chapter is to explore the various governmental agencies and programs available.

OVERVIEW

Federal, state, and local governments provide a wide range of support for soil scientists and managers. Such resources are often underutilized by the public, but these opportunities should not be overlooked. Agencies like the Cooperative Extension Service, U.S. Department of Agriculture, Natural Resources Conservation Service, Agricultural Research Service, and various other agencies are increasingly accessible via the Internet. Information can be obtained at the touch of a button from most of these services. Governmental agencies tend to maintain dependable Web sites with quality information available to anyone with access to the Internet.

Not only can soil scientists access information, but they can also interact directly with regulators and researchers working with soil. Individuals can obtain recommendations and guidance specific to their situations by communicating with the worldwide community of soil professionals. Remote access to these valuable tools and resources will become increasingly important to soil managers. Implementation will become the more limiting factor in soil management rather than lack of information.

ACTIVITIES

The activities in this chapter are designed to familiarize you with the types of governmental agencies that relate to soil management.

Equipment and Materials

> Internet access
>
> e-mail account
>
> soil sample with lab analyses

Activity

1. *Accessing soil information on the Internet.* Direct your browser to an Internet search site (e.g., Google). Enter the following search terms: "soil," "United States Department of Agriculture,"

"Agriculture," "Natural Resources Conservation Service," "LEED," and "Agricultural Research Service." Develop a favorites list of government Web sites available.

2. *Interact with soil professionals.* Develop a list of questions specific to the soil sample provided by your instructor and submit these questions to a soil professional. This could be accomplished via the Internet or face to face at a local office. While electronic communication is quicker and easier, much is to be gained through a face-to-face interaction with a professional soil scientist.

3. *Develop a lab resource list.* Find soil labs that serve your area. Develop a list of these labs, including the types of tests performed, fees, and contact information. Visit a soil-testing lab as a field trip with your class.

CHALLENGE QUESTIONS

1. What types of soil management programs are available to agriculturists in your area?

2. How are governmental agencies funded?

3. How are soil interests communicated to legislators?

4. How will we combine local soil regulations and management with global soil regulation and management? Is this desirable?

CPSIA information can be obtained
at www.ICGtesting.com
Printed in the USA
BVHW011914071121
620954BV00002B/1